How Reason Almost Lost Its Mind

How Reason Almost Lost Its Mind

The Strange Career of Cold War Rationality

PAUL ERICKSON,
JUDY L. KLEIN,
LORRAINE DASTON,
REBECCA LEMOV,
THOMAS STURM,
AND MICHAEL D. GORDIN

The University of Chicago Press
Chicago and London

The University of Chicago Press, Chicago 60637
The University of Chicago Press, Ltd., London
© 2013 by The University of Chicago
All rights reserved. Published 2013.
Paperback edition 2015
Printed in the United States of America

24 23 22 21 20 19 18 17 16 15   3 4 5 6 7

ISBN-13: 978-0-226-04663-1 (cloth)
ISBN-13: 978-0-226-32415-9 (paper)
ISBN-13: 978-0-226-04677-8 (e-book)
DOI: 10.7208/chicago/9780226046778.001.0001

Library of Congress Cataloging-in-Publication Data

Erickson, Paul, 1976–
    How reason almost lost its mind : the strange career of Cold War
rationality / Paul Erickson, Judy L. Klein, Lorraine Daston, Rebecca Lemov,
Thomas Sturm, and Michael D. Gordin.
    pages; cm
    Includes bibliographical references and index.
    ISBN 978-0-226-04663-1 (cloth : alk. paper)—ISBN 978-0-226-04677-8
(e-book) 1. Cold War. 2. World politics—1945–1989. 3. Cold War—
Philosophy. 4. Reason—Political aspects. 5. Rationalism—Political
aspects. 6. Game theory—Political aspects. I. Klein, Judy L., 1951– II.
Daston, Lorraine, 1951– III. Lemov, Rebecca M. (Rebecca Maura)
IV. Sturm, Thomas, 1967– V. Gordin, Michael D. VI. Title.
    D843.E69 2013
    909.82'5—dc23

                                                              2013013425

♾ This paper meets the requirements of ANSI/NISO Z39.48-1992
(Permanence of Paper).

CONTENTS

PREFACE AND ACKNOWLEDGMENTS

This book began at "The Strangelovian Sciences" workshop, held at the Max Planck Institute for the History of Science, Berlin (MPIWG) in March 2010. Out of that workshop a Working Group of six crystallized, who met once again in Berlin for six weeks in the summer of 2010 to write, discuss, revise, discuss again, and revise yet one more time in order to produce a jointly authored book. Our conversations, both formal and informal, were wide ranging, critical, unpredictable, sometimes heated, and always engrossing. Without them, this book could not have come into being, no matter how diligently each of us worked in solitude. We regard it as a collective work. An impeccably rational device ordered the authors' names: a randomizing computer program.

Since the summer of 2010, the manuscript has been substantially revised in light of the comments we received from readers for the University of Chicago Press. We thank Hunter Heyck and two anonymous referees for their suggestions and criticisms, which have greatly improved the book. Gil Skillman from the Department of Economics at Wesleyan University was kind enough to read and comment on the sections dealing with game theory. Karen Merikangas Darling, our editor at the Press, shepherded us through the long process from manuscript to book with patience, encouragement, and sage counsel.

Like other MPIWG Working Groups, the authors of this volume are gratefully indebted to the institute's hospitality and support, especially that of the library and Josephine Fenger, who heroically rounded up the images, sought permissions, and compiled the bibliography. Thomas Sturm's and Judy Klein's participation was supported in part by the Spanish Ministry for Science and Innovation (reference number FFI 2008–01559/FISO, to T. S.) and the Institute for New Economic Thinking (grant number

IN011–00054, to J. K.), respectively. We also thank all the participants in the March workshop, whose papers and comments proved invaluable for the conceptualization of the volume. Paul Erickson, Judy Klein, Rebecca Lemov, and Thomas Sturm are deeply grateful for the initiative that Lorraine Daston and Michael Gordin took in conceiving and coordinating our exploration of Cold War rationality.

Six weeks of the summer can be a long time to be away from home, and we all greatly appreciate the indulgence of friends and families in allowing us to work together so intensively. Finally, we acknowledge with thanks the help rendered at a crucial moment by Ivy, who pushed the button.

The Struggle over Cold War Rationality

In the Pentagon War Room, flanked by the Joint Chiefs of Staff and the Soviet ambassador, the President of the United States speaks to the Soviet premier on the special link to Moscow. They have only minutes to avert a world-destroying thermonuclear war triggered by a rogue American bomber. Everything, literally everything, depends on their remaining rational under crushing stress:

> [Ambassador] Zorubin looked carefully at the President. He realized instantly that the President had reached the brink. He could be pushed no further. Zorubin sensed that in the final analysis the President would not now hesitate to take action. Once again, the fate of the world was trembling in the balance.[1]

Economist and strategist Thomas Schelling reassured the alarmed public that such scenarios—here taken from the novel *Red Alert* (1958), most famously the loose pretext for Stanley Kubrick's nuclear farce *Dr. Strangelove, or How I Learned to Stop Worrying and Love the Bomb* (1964)[2]—were highly unlikely, but admitted that this one surpassed "in thoughtfulness any nonfiction available on how war might start." Nonetheless, for Schelling the assorted nuclear novels and films remained fantasies, no matter how ingenious. Accidents only brought the world to the brink because human beings had made choices that enabled those accidents to spiral out of control: "The point is that accidents do not cause war. *Decisions* cause war."[3] It was of utmost importance that those prior decisions be rational. In the fictional scenarios, it was the quirky human factor—something that had not been taken into account in the process of routinizing the machinery of nuclear deterrence—that unleashed the forces of destruction; and, if there was a happy ending, reined them in again. But analysts like Schelling took the opposite tack. They sought ever more reliable *rational* safeguards to tame the thermonuclear arsenals ordered by the politicians, built by the physicists and engineers, and tended by the generals. Cold War rationality in all its variants was summoned into being in order to tame the terrors of decisions too consequential to be left to human reason alone, traditionally understood as mindful deliberation.

In that implied gap between reason and rationality lay the novelty of Cold War rationality. Philosophers have debated the nature of reason— and of rationality—for millennia. There was nothing new about squabbles on that score. But the two terms had either been used as rough synonyms or had each been assigned its own domain: reason referred to the highest intellectual faculty with the most general applications, from physics to politics to ethics; rationality referred more narrowly to the fitting of means to ends (sometimes called instrumental reason) and was especially associated with economics and engineering. What was distinctive about Cold War rationality was the expansion of the domain of rationality at the expense of that of reason, asserting its claims in the loftiest realms of political decision making and scientific method—and sometimes not only in competition with but in downright opposition to reason, reasonableness, and common sense.

0.1. Rationality Enlists

By its own lights, Cold War rationality never existed. Not because the whole idea of the Cold War was irrational, in the way that the acronym for mutual assured destruction (MAD) seemed to advertise, and not because Cold War strategists were persuaded that, in the end, irrationality—fear, mischief, miscalculation, arrogance, lunacy—would likely prevail. No, the phrase "Cold War rationality" would have struck its proponents as bizarre because it sounds confined to a particular geopolitical predicament of the late twentieth century. Their aims were grander: to articulate a pure rationality, valid independently of the problems to which it was applied, and therefore also valid for everyone and always.

The aim of this book is to make the label "Cold War rationality" stick. Although one can find many of the elements that composed this form of rationality earlier, and even, arguably, advocates who assembled some or all of them into a package, it was in the United States at the height of the Cold War, roughly between the bombing of Hiroshima and Nagasaki in August 1945 and the early 1980s, that this project of articulating a particular form of reasoning commanded the attention of sharp minds, powerful politicians, wealthy foundations, and top military brass. Its home was the human sciences (variously grouped and subgrouped as the social or behavioral sciences, flexible terms with many competing definitions[4])—political science, economics, sociology, psychology, and anthropology—but with key contributions from mathematicians, statisticians, biologists, philosophers and computer scientists. The assorted theorists, policy wonks, and other figures who float through these pages did not sign up to a complete set of tenets or doctrines, a kind of thirty-nine articles of rationality. Rather, they enlisted in an intellectual campaign to figure out what rationality should mean and how it could be deployed in a world perceived to be imperiled as never before in human history.

This book is as much about their debates as about their doctrines: Where to draw the line between rationality and irrationality? Between rationality and reason? Who was the ideal bearer of rationality, the individual or the collective—or perhaps nonhumans, like animals and computers? Were empathy and emotion the friends or foes of rationality? Could situations be engineered to make people more or less rational? What methods would achieve rationality within the human sciences themselves? Above all, how could rational decision making be guaranteed when the stakes were highest and the pressures of the moment were least conducive to calm deliberation, on the brink of nuclear war? The traditional forms of practical reason and statecraft, which emphasized prudence, experience, deliberation, and consultation, seemed inadequate to the challenge, as outmoded as conventional weapons in comparison with nuclear arsenals. However abstract and technical discussions over game theoretic matrices or information-processing operations might have been, the quandaries of the nuclear age were never far from mind, as the examples in this book testify. It was first and foremost a sense of unprecedented urgency that distinguished debates over rationality during the Cold War from those over similar issues waged before and after: in the minds of the participants, nothing less than the fate of humanity hinged on the answers to these questions.

So what was Cold War rationality? An ideal type might be constructed. First of all, this rationality should be formal, and therefore largely independent of personality or context. It frequently took the form of algorithms—rigid rules that determine unique solutions—which were moreover supposed to provide optimal solutions to given problems, or delineate the most efficient means toward certain given goals (taken, in this instance, for granted). Second, complex tasks and episodes were analyzed into simple, sequential steps; the peculiarities of context, whether historical or cultural, gave way to across-the-board generalizations; analysis took precedence over synthesis. And finally, at least ideally, advocates hoped that the rules could be applied mechanically: computers might reason better than human minds. This ideal type shows the marks of its historical origins, as we will see in the ensuing chapters: on the one hand, in the mathematics of algorithms, linear programming, and game theory; on the other, in the theory and practice of economic rationalization.

Yet, as with most ideal types, Cold War rationality was rarely found in pure form. And as is also the case with even the most militantly antihistorical ideal types, such as *Homo economicus*,[5] this one too was molded by time and place. Most of our attention will be focused not on the passionate minority of true believers but on those who confronted the assumptions of this form of rationality and offered critiques and reformulations—all of them aspiring to an improved, more truly rational rationality. The more sharply the ideal was pronounced, the more explicitly it was deployed in theories of rational choice, economic optimization, or computation of algorithms on machines, the more easily its critics, both friendly and hostile, could point out its aporias and paradoxes—paradoxes now familiar to everyone who studies probability or decision theory.[6] What looks in retrospect like a loose and somewhat motley conglomerate of game theory, nuclear strategy, operations research, Bayesian decision theory, systems analysis, rational choice theory, and experimental social psychology then defined the field of contestation about what rationality should be under the radically altered conditions of the Cold War. Just because so many disciplines and approaches were involved in the debate over Cold War rationality, the debate itself nurtured hopes that the notoriously balkanized human sciences might finally rally to one banner.

Questions about the applications of rationality to the Cold War predicament suggested if not the answers then at least the form those answers were expected to take. What were the best rules of judgment and decision making for actors who wished to be rational? Could one state rules that might be implemented by computers free of the inconsistent foibles of human minds? Could the rules be given an axiomatic structure and applied to various domains in a determinate fashion? Which theories of rationality could be invoked to *explain* human behavior, especially in the domain of international relations, war, and nuclear strategy? And could one apply these theories also for the *resolution* of such political dilemmas? If so, should one? These were questions underlying a manifold of attempts to develop theories of rationality in the Cold War—attempts that were often closely related to but equally often competed with one another, like members of a family in both Freud's and Wittgenstein's senses.

The competition among these attempts will be discussed at length in the chapters that follow; for now, consider what it means that these questions and their attendant solutions were "closely related," for both proponents and opponents judged this to be the case. Affinities included an individualistic (if not egoistic) and often agonistic perspective (most obviously in game theory and nuclear strategy, but operations research also imbibed some of the competitive drive of the market); a tendency to radically simplify complex situations, abstracting from personalities and politics (whether in contrived social science "situations" or decision theory); a preference for breaking down the solutions to knotty problems into a series of steps that minimized reliance on personal skill and discretion (whether in the moves of games, the recording of social science observations, or the task schedules of operations research); and a near obsession with methods, especially algorithmic and formal ones.

The distinctive combination of stripped-down formalism, economic calculation, optimization, analogical reasoning from experimental microcosms, and towering ambitions that characterized Cold War rationality bore the stamp of an equally distinctive moment in the history of the American human sciences. Lavish government funding, new institutions like the RAND Corporation that straddled university, military, and industry settings, networks woven by select summer schools and conferences, and the sense of urgency created by the threat of catastrophic war or even human annihilation—these factors set the stage for the debates over Cold War rationality. Both the advocates of this formalistic interpretation of rationality and many of their more numerous critics circulated on the same professional tracks, spoke mutually intelligible idioms, and defined a ground of struggle that was shared as much as it was contested. Everyone was after the Holy Grail of *real* rationality; no one was willing to surrender that honorific, no matter how its definition mutated. In their analogies and debates, they slid effortlessly between, on the one hand, the nuclear standoff between the United States and the Soviet Union and, on the other, chariot races in the *Iliad*, loud-mouthed colleagues at colloquia, and children's temper tantrums—all of these were understood to be part of the same conversation, the most important conversation one could be having about what intellectuals could accomplish at that historical moment.

As the United States and the Soviet Union attained sufficient nuclear weaponry and delivery capacity to destroy each other and much of the rest

Figure 0.1. "Kennedy and Khruschev: 'OK, Mr President, let's talk!'"
(Cartoon by Leslie Illingworth, 1962; *Daily Mail* / Solo Syndication.)

of the world (that is, around 1960), strategists widened their view to absorb human interactions as well as the megatonnage of bombs and the trajectory of missiles. The world on the brink became personalized, "a war of nerves, like the wrestling bout on the brink of the cliff in so many old Western movies."[7] Within the framework of Cold War rationality, the hotheaded cowboys became cold-blooded calculators, intelligent, implacable, and symmetric (figure 0.1).

And then the perceived symmetry between the combatants began to fray, and Cold War rationality no longer seemed to be the center of everyone's frame of reference. From the middle of the 1970s, as the advent of President Richard Nixon's détente took some of the intense terror out of the nuclear standoff, and especially with the ignominious collapse of the American war in Vietnam, much of the bloom came off the rose with respect to the calculating intellectuals of the 1960s, who had made grand promises for their new rationality. Even within a more narrowly academic and disciplinary context, failure to solve problems thrown up by the theories themselves and to achieve consensus after decades of research and discussion sapped confidence that rationality could be crisply defined, much less mobilized. In the 1980s and 1990s, the elements of Cold War rationality did not wither away—far from it—yet the forces that had earlier held together the different disciplines, techniques, and policies in the crucible of the Cold War began to slacken as the Cold War itself was redefined by the proliferation of nuclear weapons and the miring of the superpowers in regional wars. The debate over rationality did not cease, as is obvious in the still-ongoing rows over rational choice theory in political science, the heuristics-and-biases program in psychology, and in many areas of ethics and epistemology. But this is now one skirmish among many, not the hottest game in town. Defining rationality was no longer the sacred quest that would unify the human sciences and keep the world safe from nuclear Armageddon.

0.2. The History of Postwar Reason

The Cold War thus furthered the growth of theories and debates about rationality, and the era can be viewed as a chapter in the long history of reason. Of course, it is a daunting task even to sketch the place of these theories and debates in the broader history of which they are a part. For example, the terms in various languages occasionally mean quite different things, due to the complicated histories of their Greek and Latin ancestors, "logos" and "ratio."[8] If you try to get at the concepts in some other fashion, you encounter additional problems.

Consider first the branching that is observable in the relevant terms in several European languages: *reason/rationality, Vernunft/Rationalität, raison/rationalité, ragione/razionalità, razum/ratsional'nost'.* In each pair, the first term is older. Are there conceptual differences behind the terminological ones and, if so, which ones? Moral philosopher John Rawls suggested one currently influential attempt to contrast the "rational" and the "reasonable," in ways that retrace some of the contours of the debates over Cold War rationality. He understood the first as referring to theories of rational choice, with a strong bias toward understanding rationality in an instrumental way. "Reason," in contrast, implied the use of moral considerations concerning the validity of goals or purposes. Rawls maintained that the reasonable could not be reduced to the rational. You could be unreasonable without being irrational.[9]

Another muddle involves the frequently unclear relation between descriptive and prescriptive aspects in theories of rationality. At the 1964 Berkeley conference "Strategic Interaction and Conflict," Schelling cast about for a less rubbery, more descriptive term: "It would be useful if rationality were not a loaded term which implies it's better to be rational, or that people who are rational are socially desirable, so they can't be eccentric or crazy. If we could Latinize the term so that one word means 'economic theory of rationality' and another means something else, we'd be better off."[10] Schelling assumed that confining the expression "rational" to the economist's meaning would give it a less value-laden, more calculating meaning—a move that came under severe attack only a few years later in controversies among psychologists and philosophers over the heuristics-and-biases research program. The concept of rationality continued to waver between descriptive and prescriptive meanings (it does so to this day). Protean in its forms and imperialistic in its contents, rationality could not be corralled so easily.

In keeping with this shape-shifting history, Cold War rationality itself had multiple elements and versions. Most of the participants in the debates over Cold War rationality subscribed to only one of its many flavors. This heterogeneity makes sense, for none of these component features inexorably summoned any of the others. Economic rationalization does not necessarily lead to algorithmic procedures, any more than game theoretical matrices intrinsically favor Bayesian statistics or even highly formalized war games. Nuclear strategists often laced intricate calculations with pop psychology. Proponents of the social science "situation" may have attempted to mechanize observation, but they did not aspire to mathematical models. These domains did not "belong together," if by that one expects there to be a rigid logic linking A to B to C.

There was no universal agreement about how to define Cold War rationality, just as there had been none about earlier conceptions of reason. Nonetheless, at least since the Enlightenment contemporaries and later thinkers had picked out particular features as central and stable that can be used here to sketch another ideal type. Traditionally, reason was seen as the highest of the mental faculties and as drawing on all of them (understanding, memory, judgment, imagination). Reasoning may be formal and even algorithmic, as in the case of algebra or logic, or substantive, as in the quintessential expressions of reason in mathematics: the demonstrations of Euclidean geometry. About such points, theorists in the Cold War would not quibble much; other differences ran deeper. To begin, the idea that machines might reason better than human minds was alien to Enlightenment thinkers. They viewed mindfulness as essential to reason in at least two ways. First, because judgments, inferences, and decisions can be right or wrong, they must be checked consciously: reason follows laws deliberately, rather than being simply subject to them. Machines cannot do this (neither, according to older conventional wisdom, could other animals). Second, the deliberations of reason encompass both complexity and contingency, the latter a particularly stubborn obstacle to automation.

Although rationality was also often taken as normative, it no longer discriminated among humans, animals, and machines—in fact, machines might outdo humans in executing algorithms. The hindrances to reason/rationality also shifted during the Cold War. Traditionally, the passions, fantasy, sloppy thinking (e.g., fallacies), ignorance, superstition, madness, rote, and self-deception had figured prominently. Debates over Cold War rationality focused additionally on inconsistency, incomputability, indeterminacy, paradox, the unexpected but crucial detail, and what has often been tellingly called "the human factor." Rationality was seen as compatible with both a certain kind of subjectivity (as in utility theory) and uncertainty (the probabilities of Bayesian decision theory), but not with inconsistency (e.g., violations of transitivity of preferences) and indeterminate solutions (e.g., n-person non–zero-sum games) or ad hoc adjustments to complexity and contingency. Also characteristic of Cold War rationality was a focus on individuals' choices and preferences—wherever these came from and whether or not they were reasonable.[11] Further, judgment, in the traditional sense of an assessment of the particulars of a case in light of universal directives (as in a case before a law court), is often in tension with rationality, which seeks to reduce complexity, either by stripping away all but the essential elements of a problem (as in a mathematical model) or by shrinking the issue to dimensions small enough to be observed under controlled circumstances (as in a laboratory experiment).

Fiery debates raged around Enlightenment reason and Cold War rationality. Like all truly interesting debates, the combatants shared assumptions, goals, and ways of arguing that allowed them to join battle on the same field. Most importantly, all were intensely committed to the view that the debate *mattered*: not just among intellectuals and within certain disciplines but to the conduct of human affairs. It is the contention of this book that Cold War rationality, like the disciplinary alliances fostered among the human sciences in the United States under Cold War conditions, cohered under the pressure of context to become such an electrified forum of contestation. The implicit rules of debate—including what could and could not count as a valid element or approach to the rationality in question—fit together at a particular point in time, and began to disentangle at another. The crucial point in what follows is that they *did* in fact cohere and shape the terms of debate, if only for a few decades, albeit with echoes that still reverberate in one or another discipline or policy specialty.

To regard mind-numbing abstractions like "reason" and "rationality" as debates situated in a time and a place rather than eternal glassy essences makes it easier to conceive of them as having histories. But what kind of history can we offer for this protean set of debates? If ever a topic cried out for intellectual history—not a timeless, placeless history of ideas, but a situated history of intellectuals and their most obsessive projects—it is surely rationality. Several of the concepts that emerged as the signature elements of what we term Cold War rationality could and did surface in other times and other milieux, but we contend that understanding context is key to explaining, first, why these ideas briefly gelled into a powerful vision for the human sciences as a whole when and where they did; second, why they were so highly valorized despite clashes with long entrenched ideals of reason and reasonableness; and third, how they came to be attacked by those who understood rationality differently but agreed on its burning importance. Our aim is to not to offer a genealogy of each of the individual ideas ("utility function," "minimax solutions," "Bayes's theorem"), which for the most part has already been admirably done within disciplinary histories.[12] Nor is it to provide a comprehensive account of Cold War rationality in its myriad scientific and policy manifestations—subject matter aplenty for a whole series of books. We want instead to identify and describe a phenomenon that for a few decades spanned several disciplines—and is therefore only partly visible within each. The scale, therefore, must be larger and the resolution less fine-grained. Most of all, we seek to explain why assumptions, applications, and vaulting ambitions that in retrospect seem barely credible, if not downright bizarre, engaged the brightest minds, attracted princely funding, and persuaded generals and politicians alike of their utility, feasibility, even necessity.[13] Most of the components of what we have dubbed Cold War rationality did not originate in the Cold War, but it was the Cold War that consolidated and glamorized them.

0.3. Braver Newer World

The same moment also saw the consolidation and glamorization of the group of intellectuals who articulated and promoted Cold War rationality. In a 1967 *Life* magazine series of three articles, Theodore H. White, historian of the Kennedy presidency, gushed over a "new priesthood, unique to this country and this time, of American action intellectuals," men (they were all men—real men, too, "husky, wiry, physically attractive men who, by and large, are married to exceptionally pretty women") liberated from musty studies and fraying tweed as they literally jet-setted around the nation for consultations and conferences. "Their ideas are the drive wheels of the Great Society: shaping our defense, guiding our foreign policy, redesigning our cities, reorganizing our schools, deciding what our dollar

Figure 0.2. Charles Hitch, head of RAND's Economic Division (1948–1961)
and thereafter assistant secretary of state (1961–1965). (Photograph by John Lonegard for
Theodore H. White, "The Action Intellectuals," *Life*, June 9, 1967, 43–76, 48.)

is worth." White described how this "silent club" of academics circulated
among college campuses, foundation headquarters, "think factories," and
government offices (figure 0.2).[14] There was a large overlap of names and
dockets between White's "action intellectuals" and what *Business Week* in
1963 dubbed "the defense intellectuals": "Whether a thinker is connected
with institutions such as Harvard, the Council on Foreign Relations, or
RAND will help determine whether his ideas get a hearing where it matters most—at the White House, Pentagon, or State Dept."[15] These forums

brought together representatives of different disciplines—and it was essential that they maintained one foot firmly in their individual disciplines—to discuss what increasingly came to be the shared project of defining and debating Cold War rationality.

To get a sense of this "silent club"—its sites, its members, and their interactions—let us briefly revisit that 1964 Berkeley conference, where we left Schelling groping for a more neutral word for the kind of rationality he and his colleagues studied. The venue was characteristically hybrid—a university campus but at a center explicitly conceived to address Cold War issues of "international security," and funded to match.[16] Almost all of the participants knew one another and were on a first-name basis, although they hailed from different disciplines and institutions and sometimes were embroiled in fierce controversies with one another. Psychologist Morton Deutsch from Columbia University, economist Daniel Ellsberg from the RAND Corporation, mathematical psychologist Anatol Rapoport from the University of Michigan, sociologist Erving Goffman from the University of California at Berkeley, game theorist Martin Shubik from Yale University, political scientist Albert Wohlstetter from the University of Chicago—and of course Schelling from Harvard—were all there, discussing, debating, joking, haranguing.[17] Taken together, they covered the waterfront in the human sciences, from psychology to economics to political science, and represented approaches ranging from the most technical (Ellsberg and Shubik) to the most ethnographic (Goffman). Yet they were all intensely engaged in discussions on the concept of rationality and the vocabulary of basic moves in game theory, illustrated with examples that ran the gamut from gambling in Las Vegas to the deployment of Strategic Air Command bombers. The people and the place (as well as the topics and the combative, chummy sociability) open a window onto the more intimate context of Cold War rationality.

First, the people: consider Schelling himself, a versatile, incisive, and quotable analyst whom we will encounter often in the following chapters. He was awarded the Nobel Prize in Economics in 2005, served as president of the American Economics Association in 1991, and taught in economics departments across the United States. But to call him simply an "economist" is to miss the point—he was also an influential analyst of international relations, nuclear strategy, disarmament, and racial desegregation policies. Born in California in 1921, he received his PhD in economics from Harvard University at the age of twenty-seven, whereupon he was awarded a position at and then left the Society of Fellows at that university to work on the board of the Marshall Plan for several years before becoming a policy advisor to the Truman administration. He then returned to academia at Yale University in 1953, where he began serious research on bargaining and negotiations, culminating in his landmark *Strategy of Conflict* (1960), a book that connected game theory to problems of decision making and strategy and was read widely across the human sciences. He wrote much of the text while on leave in London, where he spoke to several retired military officers interested in theories of limited war. When he returned from England he spent some time at the RAND Corporation in Santa Monica, California, before decamping to a position at Harvard, where he remained for thirty-one years—although without cutting his ties to government, for he consulted with the Kennedy administration on matters of national security.[18]

At RAND, Schelling joined another major figure in the world of nuclear strategy, Albert Wohlstetter. Born in New York City in 1913, Wohlstetter studied at City College and Columbia University through the Great Depression, moving on to work at the War Production Board during World War II. From 1951 to 1963 he was at RAND in Santa Monica, where he eventually rose to the level of senior policy analyst. He was almost certainly the most influential nuclear strategist of the Cold War, devoting his attention to issues of nuclear posture and deterrence, most famously in his November–December 1958 policy paper for RAND, "The Delicate Balance of Terror," which revised nuclear strategy for a post-Sputnik age.[19] At RAND, Wohlstetter was located at one of the central nodes of Cold War rationality, and he attended conferences with figures from other disciplines who shared similar preoccupations. He also consulted with the Kennedy Administration (especially during the Cuban Missile Crisis), and advised Democratic and Republican administrations thereafter. He taught political science at the University of Chicago from 1964 to 1980, where he influenced generations of future public servants who would populate the corridors of the Pentagon and State Department.

Many figures with similarly zigzagging vitae are discussed in detail in the chapters that follow (such as George Dantzig, Herbert Simon, Oskar Morgenstern, Herman Kahn, Robert Freed Bales, Anatol Rapoport, and many others), and one could add dozens more to their number. But for the moment consider the cases of just these two. Schelling and Wohlstetter came at problems of strategy from very different perspectives, and a disciplinary history of game theory in economics or theories of nuclear posture would not necessarily put them in the same room. Nonetheless, they were part of a community of sorts, reading and citing each other's publications, attending the same conferences, worrying the same bones of contention.

Their multiple, sustained interactions mapped the terrain on which Cold War rationality was attacked and defended. They did not always agree, but they saw their disagreements as pushing forward a common project. Theirs was a world of conferences like the one at Berkeley, but it was also a world of stable nodes, institutions where these individuals met face-to-face, conversed, and argued (at times bitterly, as Wohlstetter and Rapoport did at Berkeley over the morality of assumptions guiding theories of nuclear strategy, whatever their rationality[20]).

Second, the places: no account of this world would be complete without the RAND Corporation (short for "research and development"), a think tank originally established with US Air Force funds to bring precisely this kind of talent into conversation. They played the same game, but not always on the same team. For example, a popular pastime at RAND in those years was *Kriegspiel*, a variant of chess where neither player could see his (again, pretty much always "his") opponent's pieces, and a referee enumerated allowable moves. As Richard Bellman, a RAND mathematician, recalled, "Kriegspiel is a good game since it combines various features of chess and poker," and he himself "got involved in simulation through kriegspiel," which often lasted well beyond a lunchtime diversion.[21] It was a fitting activity: a mixture of ancient strategy and manly bluffing. In German, of course, the name means "war game," and that was what RAND was about. It was organized into departments and favored informal modes of socializing modeled on university campuses—from which many of its fellows hailed and to which many of them would return, often on a revolving-door basis. In the relaxed environment of Southern California, just a short jog from the beach (not that too many action intellectuals were surfing men), discussions of nuclear strategy and logistical planning drew from political science, economics, sociology, and mathematics. The bonds formed at and through RAND reverberated within Cold War rationality and were part of what kept the discussion going.[22]

But reducing the story to one institution would be as myopic as constraining it to one discipline, for the Cold War rationality debates could be found farther afield, and at various points on the political spectrum. For example, the Center for Advanced Study in the Behavioral Sciences (CASBS) at Stanford University, south of San Francisco, attracted many psychologists of a less hawkish persuasion than your typical RAND fellow, but the same tools (optimization, Bayesian statistics, game theory) were to be found in its offices as well. The Cowles Commission for Research in Economics—founded in Colorado Springs in 1932 then moved to the University of Chicago in 1939 and to Yale University in 1955—was another forum that brought these action intellectuals together, this time to link mathematics and statistics to economic theory. The Council on Foreign Relations in New York (and also Washington DC) focused another set of minds on the problems of international conflict and stability. All of these sites would concentrate thinkers, but then send them back home to their book-lined, pipe-smoke-filled offices at universities, foundations, and think tanks—and also to their home disciplines. This intellectual catch-and-release constituted White's "silent club" of action intellectuals and encouraged the unifying aspirations of Cold War rationality.

Third, what was discussed and how it was discussed: Although the obsessive focus of the discussions—in Boston and in Berkeley, at RAND and the CASBS, in suited-up Washington meetings and shirtsleeves conferences—were Cold War issues, the conversation could careen between nuclear Armageddon and unhappy marriages. Game theory was tried out to model the East–West standoff in occupied Berlin and for when to take the Fifth Amendment, sometimes in the space of a few minutes (if the transcripts of the Berkeley conference are anything to go by). Cold War applications ratcheted up the stakes of these discussions, and what at first (and sometimes second) glance looks like a propensity for wild analogies between the apocalyptic and the prosaic can also be seen as an expression of the drive to generalize across a wide range of phenomena typically parceled out among different disciplines. The atmosphere of these discussions, conducted with brio and brilliance, mingling technical insights with homespun anecdotes, informal but in dead earnest, also united the discussants in a shared sociability, even if—especially if—they were almost never united in their views. White joked that the action intellectuals he interviewed never stopped talking, were always gabbing and sparring with one another in each other's offices, in the halls, around the seminar table. "The trouble is that every brilliant proposal runs into an equally brilliant counter-argument. If all of these guys could agree with one another, it would be a wonderful country."[23]

White nonetheless insisted that the aggressively gregarious action intellectuals were serving their country in times of dizzying change and mind-boggling peril. It was not only the masses—who either (depending on which pundit one read) had been whipped up into a murderous frenzy by demagogues or stultified by too much television[24]—who needed expert guidance. So did the sober, well-educated leaders of the nation. An effusive White House admirer of Charles Hitch (figure 0.2), declared, "This is the first time that someone is seriously sitting down, taking an analytical approach, and making a rational allocation of resources to buy defense."

As Hitch himself put it in the same article, Cold War questions "require an analytical approach, an ability to think in abstract or conceptual terms. This is the sort of thing an intellectual, by virtue of his training and mental discipline, can do better than a military professional who is not an intellectual."[25] White concurred: "As the world outruns its comprehension of itself, inherited tradition no longer grips onto reality. . . . Thus, with almost primitive faith, American government has turned to the priesthood of action intellectuals—the men who believe they know what change is doing, and who suggest that they can chart the future."[26] Many of the Cold War rationalists shared White's views. The hurtling pace of technological advances, especially but not exclusively in the realm of weaponry, had, they believed, overwhelmed the "Stone Age mentality" of so-called *Homo sapiens*, a mindset that had shown itself unable to manage automobiles wisely, much less nuclear bombs (figure 0.3). Princeton political economist Oskar Morgenstern, coauthor with mathematician John von Neumann of the seminal treatise *Theory of Games and Economic Behavior* (1944),[27] warned that no political system or human science had yet been able to deal with "the difficulty of making 'rational' decisions in the face of an ever changing world."[28] Political scientist and erstwhile US ambassador to the Soviet Union George Kennan agreed and feared that democracy would be unable to cope with the torrent of new technologies: "In a complex technological environment, the common man cannot know what is really good for himself."[29]

On this account, technology had intensified human capacities for destruction, while simultaneously accelerating the pace of decision making

Figure 0.3. Bone-wielding ape uses the first tool to smash a skeleten, linking the origins of technology to violence. (Still from *2001: A Space Odyssey* [1968], directed by Stanley Kubrick.)

beyond humans' ability to reason effectively. Stone Age man had been well versed in violence, but he also wielded his stone axe on a comparatively slow time-scale. There was time for pleading, for self-defense, for evolving stratagems that took into account the complexity and contingencies of the situation. The nuclear missile, which would arrive from across the globe in thirty minutes (give or take), meant there was no time to account for details. And even if you had the time, you couldn't recall the missile. You *could* recall a bomber, but Cold War strategy inserted algorithmic rules to make that impossible as well—turning the manned bomber into an ersatz missile, staffed by reasonless beings and forming the basis for numerous Cold War quasifictions from *Red Alert* onward. Algorithmic variants of Cold War rationality needed to constrain the scope for reflective maneuvering, to protect it from Stone Age tampering, reckoned the analysts at RAND and elsewhere. Rationality potentially provided a way to think through the unthinkable—not just to render the unthinkable into a problem that could be solved through algorithmic processes but voluntarily to constrain our practices of reasoning so that we would become the kinds of subjects who refrained from incinerating the planet.

0.4. Rationality across the Curtain

From the standpoint of the American Cold War rationalists, the United States and the USSR shared both the predicament and its rational solution—just because it was rational and therefore universal. American strategists drummed in the message: America and the Soviet Union occupied symmetric positions, adversaries playing the same pieces across a shared chessboard. Novelists and filmmakers depended on the conceit: "You have the same computers we do!" shouts the American General Bogan to his Soviet counterpart in the accidental-nuclear-war film *Fail-Safe* (1964). The symmetry was all in the eye of the (American) beholder, however, for American-style Cold War rationality had no precise Soviet mirror image— nor, given its dynamic and contingent assortment of disciplines and theories, should we expect any necessary twinning.

Consider the reception of Morgenstern and Von Neumann's treatise on game theory, a mathematical approach to strategy that became emblematic, even to its critics, of Cold War rationality tout court: "If *any* theory of conflict can lay claim to rationality, it is game theory, a mode of analysis endowed with mathematical rigor and unencumbered by either practical difficulties (e.g., those of collecting data) or psychological biases."[30] Although not adopted by economists as rapidly as Morgenstern (for one) had hoped,[31] this book sparked a broader conversation about the utility of game theory across a variety of different disciplines and was translated into several other languages as the theory's impact in the American context became evident. In the United States and Great Britain at least, game theory spread from economics to political science to psychology to biology, as discussed later in this book. But the Soviet response was belated, circumscribed, and comparatively tepid.

When Morgenstern traveled to Vilnius in Soviet Lithuania to attend the Second All-Union Conference on Game Theory in June 1971, he was presented with a copy of the Russian translation of *Theory of Games and Economic Behavior*, which had been prepared by 1968, twenty-four years after its English-language debut, but was not published until 1970. The audience of 168 attendees applauded Morgenstern, but they did not take up in droves his call to expand the applications for game theory. Game theory simply did not cohere within the human sciences in the Soviet Union. Consider the frustrations of its most enthusiastic Soviet proponent, the mathematician Nikolai N. Vorob'ev (also the editor of the translation of Von Neumann and Morgenstern's text). Having begun research on this topic in the mid-1950s, he published two books on matrix games and infinite zero-sum games in 1961 and 1963, respectively, and then pushed for the organization of an all-union conference on the field. The first was in Yerevan, the capital of Soviet Armenia, in November 1968, followed three years later by the Vilnius event, and a final one in Odessa in 1974.[32] The delays between the conferences and their locations are indicators of game theory's relative marginalization in the Soviet context. These were not meetings in Moscow or Leningrad; they were relegated to the provinces.

The proceedings of the meetings reveal something else about Soviet game theory: its firm establishment as a subfield of mathematics. The *Great Soviet Encyclopedia*, in an article coauthored by Leonid Kantorovich—no mean mathematician himself—defined game theory as "an area of mathematics which has as its subject the study of mathematical problems, connected to finding the most advantageous line of behavior (optimal strategy) for each of the participants of the game."[33] Vorob'ev was equally clear about what was *not* covered by Soviet game theory:

> The theory of games is a system of normative models, i.e., it is used to determine the optimality of decisions taken in conflicts. In this regard game theory is not concerned with descriptive models that describe the processes of evolution of conflicts, or constructive models that describe the factual

implementation of decisions in conflicts, or, finally, prognostic models that predict the course and outcome of conflicts.[34]

Vorob'ev further noted, with some disappointment, that "the applied aspects of game theory in our country are developing successfully, although not as intensively as one might have expected."[35] For example, not a single philosopher attended the Vilnius events—a discouraging sign in the Soviet Union.

By contrast, philosophers and members of other disciplines in the United States were clearly smitten with game theory. So what was different for the Soviets? Part (but only part) of the explanation might be linguistic, related to the connotations and deep histories embedded in choices of words. Recall Schelling's observation about the particular prescriptive valence of "rationality" in English, a language which lacked a distinction to reflect the contrasts between what economists meant by that word and what philosophers had traditionally construed reason to be. In the case of Russian, such a distinction did indeed exist, thus acting as a partial brake on the sorts of slippages that animate the chapters that follow. "Rational" in Russian is related to ratios, as in the term "rational numbers"; it, and therefore game theory, had nothing to do with the important philosophical territory of reason (*razum*). The value-laden meaning and the economists' meaning simply could not be confused in Russian—and given the pervasiveness of economics within Cold War rationality, part of the cohesiveness of the American discourse was lacking.

There were other reasons why the human sciences that emerged in the Soviet Union were not freighted with notions of rationality. To the extent that there was a Soviet discipline that took on some of the many burdens shouldered by Cold War rationality in the West, it was cybernetics. Cybernetics—the science of feedback, communication, and control christened by American mathematician Norbert Wiener in 1948—entered the Soviet Union in 1951 under the sign of repression.[36] But cybernetics was soon rescued from ignominy in the annals of Soviet science, both as the result of support from philosophers and also (and more importantly) from mathematicians and computer scientists. Those latter two camps bivouacked much closer in the Soviet context than the American, where computers were (and to some extent, still are) understood as a province of engineering.[37] In the civilian sectors of the Soviet Union, they were assigned to mathematics: computers were dubbed "mathematical machines," and software "mathematical support." Since mathematics had an elevated status as being "pure," it helped liberate cybernetics from official ideology by endowing it with some of its own aura of objectivity, although cybernetics' tangles with philosophy kept it in the limelight while game theory mutely stood stage right.[38] In the late 1950s, the very moment that cybernetics was fragmenting in the United States into separate disciplines (chief among them artificial intelligence), it acquired tremendous resonance for its Soviet advocates. As historian Slava Gerovitch notes:

> In the United States, Wiener's original eclectic synthesis of diverse scientific and engineering concepts did not hold together; various threads of the cybernetic quilt—computing, control engineering, information theory, operations research, game theory—soon parted ways. Soviet cyberneticians, on the contrary, regarded cybernetics as the potential basis for a grand unification of human knowledge.[39]

As such, it was appropriated both by state technocrats and by dissidents as a way to think about *knowledge* in a manner distinct from the dominant Marxist philosophy of dialectical materialism.

This is a fascinating story, but it is quite different from what happened in the United States. Although on the surface, the end result might resemble rational-choice policy analysis in America, cybernetics never held as much sway in so many diverse areas (in part because some zones, like philosophy, were inoculated by the official Marxism that made cybernetics so attractive as an alternative to individuals in other fields), but also for structural reasons having to do with the organization of intellectual firepower in both countries. Much of the motivation on both sides of the Cold War for the development of mathematical analysis of behavior and efficient operations was military in origin. In the Soviet Union, the military knowledge infrastructure existed in parallel to the civilian infrastructure (centered on the Academy of Sciences), with only limited points of connection—and, as in the United States, the scale of military support greatly outweighed nonmilitary sources of funding. In this context, academic trends did not proliferate and cross-fertilize in the same way as in the United States, where civilian contractors moved in and out of military consultancies with dizzying frequency. The kind of interlinked hybridization between, for example, RAND, the US Department of Defense, and universities chronicled in the chapters that follow was not present to nearly the same degree in the Soviet Union. The Soviet case deserves a history in its own right, one that would be at least as long as the present book. In the chapters that follow, the Soviet Union remains omnipresent only as the imagined player across the chessboard, as it did for the American participants in the Cold War rationality debates.

But the Soviets do provide us with a pointed question to be posed to the American case. On the American side of the Cold War fence, if only for a few decades, operations research, game theory, strategic deterrence, linear programming, decision theory, and the experimental social sciences all seemed to be converging on similar problems with similar tools and standards for what constituted a satisfactory solution. All seemed to be part of the same project. That this did not necessarily have to be the case is evident from the Soviet counterexample; there, the elements of what we have called Cold War rationality did not cohere, the constituent parts laminated instead of alloyed. The glue that held the discourse of Cold War rationality together was contextual and, eventually, it dissolved: the meanings of both rationality and irrationality ramified; the urgency of the applications faded.

The historiography of the various components reflects the instability of this underlying alliance, and rightly so in hindsight. Rational choice theory, game theory, cybernetics, artificial intelligence, operations research, economics, military strategy, computing, cognitive science, and even philosophy of science have each been the subject of excellent studies, without which the present volume could not have been written.[40] The impulse behind these studies hits home: each was indeed its own separate field, and has evolved into a rather different animal from the rest. S. M. Amadae in particular has moved beyond even these already capacious disciplinary contexts and emphasized the emergence of rational-choice theory in several fields (economics, social theory, political science).[41] Extensive attention has also been paid to the military funding of many of these developments in Cold War science, especially in the human sciences, to pivotal institutions, such as the RAND Corporation, and to key figures, such as Herman Kahn, Herbert Simon, Morgenstern, Von Neumann, Wiener, Schelling, and others. These accounts are indispensable for understanding the who, what, where, why, and when of the various scenes of Cold War rationality, and we have drawn on them extensively.[42] But we are after a different quarry: not the emergence of a specific theory or science, or the establishment of a particular institution, or the trajectory of an influential individual, but rather a change in what it *meant* to be rational in the age of nuclear brinkmanship.

0.5. How to Read This Book

This story can be read as a continuous one of how the answers to that question—what it means to be rational—evolved and fragmented in the pressure-cooker atmosphere of the Cold War.[43] It begins with the most common element of the Cold War rationality debates: formalization and economic rationalization. Chapters 1 and 2 describe how, both in theory and in practice, these algorithmic approaches shaped new ideas of rule-governed rationality. Chapters 3, 4, 5, and 6 examine the engagement between those formalized modes of rationality and the traditional science of the human mind, psychology, or the nonhuman mind as understood by evolutionary biology. The Cold War looms large in all these accounts, whether in the form of the nuclear arms race (chapter 3), the examination of officers making split-second decisions in small groups (chapter 4), or the ever-proliferating prisoner's dilemma as the signature object of study in game theory (chapter 5). In chapter 6, psychology itself, previously a contributor to the exchanges of ideas and individuals that defined the field of Cold War rationality, began to dissolve it by redrawing the boundary between the rational and the irrational. Previous attempts to enrich rationality by mixing in practical constraints (as in Herbert Simon's "satisficing" in chapter 2) or the emotions (as in Charles Osgood's GRIT technique in chapter 3) or social dynamics (as in the "special situation" in chapter 4) or moral reasoning (as in Anatol Rapoport's analysis of prisoner's dilemma in chapter 5) had partnered psychology with the economists' assumptions of self-interested maximization. But later psychological research purported to show stubborn, widespread biases and inconsistencies in actual human reasoning. In the process, the science of psychology shouldered the responsibility for explaining deviations from rationality, leaving its definition to other fields.

The antecedents to Cold War rationality crossed mathematics and economics to create powerful hybrids that reduced first calculation, then action, and finally calculation again—but this time on a grand scale—to rules. Chapters 1 and 2 trace the arc from the calculation of logarithms during the French Revolution by human computers to the advent of linear programming and management science in the decade following World War II. Chapter 1 follows the career of rules from the Enlightenment, when they referred primarily to exemplary models or the tacit knowledge expressed by rules of thumb, to the mid-twentieth century, when rules had become first and foremost algorithms. Chapter 2 chronicles the story of mathematical programming from its embryonic deployment in the Berlin Airlift, when the US military turned to applied mathematicians and economists for help in allocating resources in military operations and blending aviation fuel. These nascent attempts at optimization with limited computational resources inspired Herbert Simon's influential distinctions between "satisficing and maximizing" and between "procedural and substantive rationality." Whether the task was making astronomical calculations or organizing runway schedules, the economic rationalization of labor into "rules of action" made possible further mechanization by algorithmic rules.

The following three chapters take up the story of Cold War rationality as it was deployed in the human sciences from the 1950s through the early 1970s. As the Americans and Soviets stumbled from nuclear crisis to proxy war to escalatory flashpoint, analysts began to wonder whether human psychology was a help or a hindrance to decision making in volatile political situations like the Cuban Missile Crisis of 1962. At the peak of Cold War tensions and terrors in the 1960s, as detailed in chapter 3, some psychologists argued that an assumed symmetric predicament and psychology of the combatants could be used to de-escalate the arms race and move back from the nuclear brink, if only "groupthink" were avoided. But for these scholars, mirror-image symmetry encompassed the nonrational (but not necessarily *ir*rational) as well as the rational, the psychologically predictable as well as the logically computable. This was the empirical price certain psychologists exacted to enable Cold War rationality to continue to grasp and understand the world.

If it could. For the real world began to seem too large, too complicated, too messy to yield to formal models and elegant theories. How could a more tractable but still realistic world be constructed? Social scientists proposed two different strategies for investigating how rationality might cope with the real world, explored in the following two chapters. Chapter 4 tells of the promotion of the "situation"—a contrived, circumscribed environment that constrained the subjects and permitted unobtrusive observation by the social scientists—and its tremendous spread out from its Harvard cocoon to studies of mother–infant bonds, college student teamwork, submarine crews, and Pacific islanders. This strategy aimed to change behavior as well as to study it: by a process of interaction and feedback, the researcher could control irrational processes and transform the group into the bearer of rationality.

There were other paths to simplification: instead of constraining the complexity of the actual environment so it could be broken down, analyzed, and even harnessed, an alternative strategy illuminated only those aspects of a problem considered relevant for rational understanding; the rest was so much detritus to be ignored. The classic exemplar of this technique was embedded in a matrix—represented in chapter 5 by the ubiquitous "prisoner's dilemma"—the emblem of game theory, a branch of mathematics that stripped human interactions down to the moves of adversaries bent on winning a game with well-defined rules. Economists, political scientists, strategists, and some psychologists flocked to this alternative world, which since the 1970s contributed to a reformulation of evolutionary theory in which even genes became rational game players. Both of these brave new worlds of Cold War rationality were soon invaded by the very empirical complexity they were supposed to eliminate.

Chapter 6 chronicles the return of the residual, the growing attention to the gap between formal standards of rationality derived largely from logic and statistics and actual human performance. In this chapter we see how for some psychologists, followers of the "heuristics and biases" school that mathematician Amos Tversky and psychologist Daniel Kahneman established in the 1970s and '80s, irrationality was so indelibly engraved in the human mind that even expert training could hardly overcome it. Despite both philosophical and empirical criticism, much of it centered on how "rationality" and "irrationality" were defined in the experiments, Kahneman and Tversky's results were and remain influential in other disciplines, including policy analysis. By emphasizing the divergence between actual human reasoning and standards of formal rationality such as logic and Bayesian statistics, they implicitly reinforced the normative authority of the latter. Whereas the earlier debate over rationality had been capacious enough to encompass various attempts to modify formal rationality with empirical content and novel techniques, the impact of heuristics-and-biases psychology was to shift attention to irrationality, understood primarily as violations of logical consistency and some (but not all) formulations of probability theory—thereby implicitly restricting rationality to these formal approaches. By effectively narrowing the field of debate over Cold War rationality, this approach thereby inadvertently undermined it.

Cold War rationality may have lost its coherence, and, in some quarters, its credibility, but its elements continue to thrive in the most diverse disciplinary contexts. Nonetheless, the apparent strangeness of Cold War rationality now, some twenty-five years later, seems to leap from the page— its obsessive narrowing of vision, its mind-bogglingly implausible assumptions, its devotion to method above content, and above all its towering ambitions. A certain laser-beam brilliance, as focused as it was intense, was yoked to fierce earnestness of purpose. Both were magnified by the exhilaration of teamwork and by the direness of the challenge. The mission was nothing less than to save the world. Under maximal pressures like these, the debates over Cold War rationality took shape as programs, worldviews, even as crusades. In retrospect, the megalomania can at times seem almost comic, out of all proportion to reasonable expectations. Yet the searching discussions about how and why to be rational *in extremis* preserve some of the most adventurous episodes in the twentieth-century life of the mind.

ONE

Enlightenment Reason, Cold War Rationality, and the Rule of Rules

Sometime in 1952, RAND Corporation mathematician Merrill Flood decided to bring work home from the office. He asked his three teenage children to bid by a reverse auction[1] for an attractive babysitting opportunity on the condition that "they would tell me how they reached their agreement." After a week of deliberation, the teenagers, who had been given permission to better their collective lot by forming coalitions, were still unable to reach an agreement. Worse, the winning individual bid of ninety cents was grossly irrational according to the game-theoretic pay-off matrix, which Flood proceeded to calculate. Flood drew sweeping conclusions from this domestic experiment: "This is probably an extreme example, although not really so extreme when you compare the magnitude of the children's error with that made by mature nations at war because of inability to split-the-difference. I have noticed very similar 'irrational' behavior in many other real life situations since August 1949 [when the Soviet Union exploded its first atomic bomb] and find it to be commonplace rather than rare."[2] Rationality began at home, but its ambitions embraced the wide world.

From teenager befuddlement to wartime stalemate: Flood and his colleagues at the Santa Monica think tank RAND[3] were after an articulation of rationality so powerful and so general that it would apply to situations as prosaic as haggling over the price of a used Buick with nuclear strategist Herman Kahn (another of Flood's homespun experiments) and as apocalyptic as nuclear warfare. Flood, who had made a name in World War II operations research (described in chapter 2) and later (as we will see in chapter 5) contributed to the formulation of the prisoner's dilemma, had by the early 1950s grown skeptical about the applicability of game theory to anything but the most trivial parlor games.[4] But neither he nor his fellow analysts at RAND and other bastions of Cold War research ever doubted that whatever rationality was, it would be a matter of rules, the more mechanical the better. Perhaps the axioms of John von Neumann's and Oskar Morgenstern's game theory would have to be modified in light of experiments like the one performed on Flood's children or the SWAP war game played by his fellow RAND researchers (figure 1.1);[5] perhaps a superego or even a neurosis would have to be programmed into the "rational mind"

Figure 1.1. RAND analysts play in earnest at war games.
(*Life*, November 5, 1959, 156.)

of a mechanical deliberator;[6] perhaps actors would have to be taught to behave "in a spirit of calmly aggressive selfishness."[7] No matter how heterodox, however, attempts to model rationality almost never questioned one precept: rationality consisted of rules.

In the two decades following World War II, human reason was reconceptualized as rationality. Philosophers, mathematicians, economists, political scientists, military strategists, computer scientists, and psychologists sought, defined, and debated new kinds of norms for "rational actors," a deliberately capacious category that included business firms, chess players, the mafia, computers, parents and children, and nuclear superpowers. Older concepts of reason had sometimes been disembodied, the property of an ethereal Christian soul or a Cartesian *res cogitans*, but new-fangled views of rationality often departed from materiality (and humanity) altogether: "The rule that such a device is to follow," explained MIT computer scientist Joseph Weizenbaum, "the law of which it is to be an embodiment, is an abstract idea. It is independent of matter, of material embodiment, in short, of everything except thought and reason."[8] As Weizenbaum himself went on to argue, the reason in question was restricted to "formal thinking, calculation, and systematic rationality."[9] What made both the generality and the immateriality of rational actors conceivable was the implicit assumption that whatever rationality was, it could be captured by a finite, well-defined set of rules to be applied unambiguously in specified settings—without recourse to the faculty of judgment so fundamental to traditional ideals of reason and reasonableness. Von Neumann and Morgenstern articulated their definition of rationality in the context of game theory thus: "We described in [section] 4.1.2 what we expect a solution—i.e., a characterization of 'rational behavior'—to consist of. This amounted to a complete set of rules of behavior in all conceivable situations. This holds equivalently for a social economy and for games."[10] The solution to the failure of extant rules in logic and arithmetic to cover the full range of decision making under uncertainty was, according to University of Chicago economist and Cowles Commission member Jacob Marschak, more such rules: "We need additional definitions and postulated rules to 'prolong' logic and arithmetic into the realm of decision. We shall define rational behavior as that which follows those rules, in addition to the rules of logic and arithmetic."[11]

And not just any kind of rules: it was above all algorithms—for centuries the exclusive province of arithmetic but extended to logic in the late nineteenth century and from logic to all of mathematics in the early twentieth century—which characterized these attempts to define rational behavior.

Algorithms did not even merit an entry in one of the most comprehensive mathematical dictionaries of the mid-nineteenth century,[12] but by the turn of the twentieth century, a flourishing research program in mathematical logic had elevated the humble algorithms of elementary calculation to the status of a model for the foundations of all mathematical demonstration.[13] In a seminal treatise, Russian mathematician A. A. Markov described the three desiderata of an algorithm: "(a) the precision of the prescription, leaving no place to arbitrariness, and its universal comprehensibility—the definiteness of the algorithm; (b) the possibility of starting out with initial data, which may vary within given limits—the generality of the algorithm; and (c) the orientation of the algorithm toward obtaining some desired result, which is indeed obtained in the end with proper initial data—the conclusiveness of the algorithm."[14] Although they often described their epoch in terms of complexity, uncertainty, and risk and conjured the specter of a nuclear war triggered by accident, misunderstanding, or lunacy, the participants in the debate over Cold War rationality believed that the crystalline definiteness, generality, and conclusiveness of the algorithm could cope with a world on the brink.

Theorists of games, strategic conflict, artificial intelligence, and cognitive science diverged frequently and substantively on major issues: for example, cognitive scientist Herbert Simon's program to model "bounded rationality" with heuristics clashed with economists' imperative to optimize;[15] economist Thomas Schelling was skeptical about the usefulness of zero-sum games for modeling strategic decisions;[16] even Morgenstern wondered whether, *pace* the adversarial assumptions of game theory, cooperation might not be "more natural" than conflict in many situations.[17] The Cold War rationalists, ever critical of themselves and each other, by no means constituted anything like a unified program, much less a school. What nonetheless justifies the label is the shared assumption, rarely examined but always fundamental, that whatever rationality was, it could be stated in algorithmic rules—whether these were strategies in game theory, the consistency specifications of personal utilities, linear programming code, actuarial formulas for clinical decisions, or cognitive representations.

What was novel about this view of rationality? After all, algorithms are as old as addition, subtraction, multiplication, and division. Visions, theories, and devices that seem to foreshadow this or that element of Cold War rationality can be found in other times and places: Gottfried Wilhelm Leibniz's seventeenth-century dream of reducing reason to a calculus; Daniel Bernoulli's eighteenth-century explorations of how mathematical expectation in probability theory could be redefined to express what economists later called utility; Charles Babbage's nineteenth-century project for an analytical engine that would perform the operations of mathematical analysis as well as those of arithmetic; William Stanley Jevons's slightly later logic piano that mechanically derived conclusions from premises.[18] In retrospect, any and all of these may look like anticipations of the rule-bound rationality pursued so energetically in the mid-twentieth century, and, as we'll see, they were sometimes enlisted in attempts to provide game theory or utility theory or artificial intelligence with eminent ancestors. But only in retrospect do these dispersed ideas and inventions hang together: for Bernoulli and other early probabilists, for example, utility theory, grounded in subjective preferences, had little to do with mechanical calculation; the moral that Babbage drew from his difference engine was not that it was artificially intelligent but rather that computation, even complex computation, required little or no intelligence. Human reason was often defined in opposition to mechanical rule following (or the rote behavior of animals). As an 1842 account of Babbage's plans for an analytical engine put it, the "mechanical branch" of mathematics must be distinguished from "the domain of understanding . . . reserving for the pure intellect that which depends on the reasoning faculties."[19] Until the middle decades of the twentieth century, algorithmic rules, most especially those executed by machines, seemed the least, not the most promising materials for a normative account of rationality.

In order to take the measure of just what was new about various versions of Cold War rationality, we must therefore step back and survey its emergence against the background of a longer history of reason and rules. Only then does the historicity of Cold War rationality snap into focus: under what circumstances could mechanical rule following, previously excluded from the "domain of understanding," become the core of rationality? This chapter traces how the elements of Cold War rationality were available (and, at least in one notable case, united) at latest by the mid-nineteenth century. But far from solidifying into a new ideal of rationality, they underwent radical intellectual (and economic) devaluation, as working to rigid rules was first handed over to badly paid laborers and eventually to machines. It was in the context of the Cold War that those same elements— algorithmic rules impervious to context and immune to discretion, rules that could be executed by any computer, human or otherwise, with "no authority to deviate from them in any detail"[20]—came together as a new form of rationality with glittering cachet in the human sciences and beyond. To tell this story in its entirety would require volumes, encompassing everything from the history of philosophy since the Enlightenment to the rise of the modern bureaucracy to the development of the computer (and perhaps also the cookbook). Here, however, we will concentrate on those features of earlier accounts of reason that seem to most resemble aspects of Cold War rationality: Enlightenment applications of arithmetic algorithms and probability; nineteenth-century attempts to mechanize calculation; and the shift in the meaning of rule from model to algorithm.

We begin with a comparison of Cold War rationality to older alternatives, especially those Enlightenment versions that seem to resemble it most closely (and which were sometimes cited by the Cold War rationalists as forerunners). Key to the contrast between Enlightenment and Cold War versions of rationality is the rise of the modern, automated algorithm in connection with the economic rationalization of calculation. Rules too have their history, and the allure of rules as the backbone of rationality demands explanation. Against this background, algorithm-driven rationality emerged as a powerful tool and seductive fantasy. Or, as some critics maintained, as a powerful fantasy and a seductive tool, for its ambitions and applications were from the outset and remain controversial. Neither the rise of mathematical logic in the first half of the twentieth century nor the spread of computers in the second suffices to explain why algorithm-centered rationality became compelling in the American human sciences during the Cold War. Even within their own ranks, the Cold War rationalists struggled to maintain the consistency and clarity of their rules in the face of phenomena such as emotional outbursts, neuroses, indecision, dissent, caprices, and other manifestations of what they came to call problems of "integration," whether on the part of world leaders or their own children. On the mock gothic campuses of leading universities, in the studied informality of think tank offices, in front of room-sized computers named ENIAC and MANIAC and in the none-too-tranquil bosom of their families, the Cold War rationalists pondered the coherence of both society and the self (figure 1.2).

1.1. "Let Us Calculate"

In December 1971 Princeton professor of political economy Oskar Morgenstern wrote to his colleague Margaret Wilson in the Philosophy Department to ask for the source of a visionary quotation from the seventeenth-century philosopher and mathematician Gottfried Wilhelm Leibniz:[21]

Figure 1.2. The studied informality of a RAND Corporation office, meant to promote intense, open-ended discussions. (*Life*, November 5, 1959.)

For all inquiries which depend on reasoning would be performed by the transposition of characters and by a kind of calculus, which would immediately facilitate the discovery of beautiful results . . . it would be easy to verify the calculation either by doing it over or by trying tests similar to that of casting out nines in arithmetic. And if someone would doubt my results, I should say to him: "Let us calculate, Sir," and thus by taking to pen and ink, we should soon settle the question.[22]

Morgenstern, German-born, Austrian-educated, and a certified *Bildungs-bürger* who strewed maxims from La Rochefoucauld and historical analogies to the military campaigns of Charles V and Napoleon in his lectures on arms control to the Council of Foreign Relations,[23] had collected materials for a history of game theory (never completed). In addition to giving his shared brainchild an intellectual pedigree, Morgenstern seems to have been wrestling with, on the one hand, the tensions between his earlier education, heavily inflected with philosophy and history as well as his own earlier reservations about the unreality of hyperrational theories used to make economic forecasts, and, on the other, the formality and simplifying assumptions of game theory.[24] He, however, was not alone among the Cold War rationalists in seeking forerunners among the Enlightenment probabilists, especially in the work of Marie Jean Antoine Nicolas de Caritat, the Marquis de Condorcet.[25] Were they right to see themselves as reviving a very distinctive form of Enlightenment reasonableness?

Certainly, Enlightenment luminaries were well acquainted with the mathematics of games and the refinements of machinery; some were moreover fascinated by the possibility of turning probability theory into a reasonable calculus and by automata that mimicked the behavior of humans and animals. For example, Condorcet once computed the minimum probability of not being falsely convicted of a crime that a citizen in a just society must be guaranteed on the analogy of a risk small enough that anyone would take it without a second thought—such as taking the packet boat from Dover to Calais in calm weather on a seaworthy boat manned by a competent crew.[26] Immanuel Kant suggested that the intensity of belief be gauged by how much the believer was willing to wager in a bet as to the conviction's truth or falsehood: "Sometimes he reveals that he is persuaded enough for one ducat but not for ten."[27] (In the same passage, Kant avowed that he himself would be willing to risk "many advantages of life" in a bet on the existence of inhabitants on at least one other planet besides Earth.) The mathematics of games seemed to these Enlightenment thinkers fertile in lessons about how to reason with exactitude and consistency.

Machines in the form of automata similarly stimulated speculation about how far the analogy between human beings and machines could be stretched, whether in the form of beguilingly realistic automata that played the flute or wrote "cogito ergo sum" with a quill pen,[28] or in that of materialist treatises like Julien Offray de la Mettrie's *L'Homme machine*

Figure 1.3. A musician automaton, ca. 1770, which pressed the
keys of the miniature organ and also appeared to breathe, fabricated
by the Swiss watchmaker Pierre Jaquet-Droz (1721-1790).
(Musee de l'art et d'histoire, Neuchâtel, Switzerland.)

(1748), which described the human body as "a machine that winds its
own strings" and asserted the soul to be "but an empty word."[29] One celebrated eighteenth-century automaton, the Turkish chess player first shown
in Leipzig in 1784, dramatized the possibilities of the "mechanized higher
faculties" (although it was ultimately exposed as a fraud).[30] There would
have been nothing to shock a well-read Enlightenment *philosophe* in musings about intelligent machines—a category that perhaps included human
beings (figure 1.3).

What would have flummoxed even the most enlightened of Enlightenment readers was the central role of algorithmic *rules* in defining rationality,
more precisely the understanding of both decision making and machines
as sets of such rules, at once purely conventional and rigidly determined.
British mathematician Alan Turing captured the difference in a seminal
1950 article on computing and the possibility of mechanical minds:

The book of rules which we have described our human computer as using
is of course a convenient fiction. Actual human computers really remem-

ber what they have got to do. If one wants to make a machine mimic the behaviour of the human computer in some complex operation one has to ask him how it is done, and then translate the answer into the form of an instruction table. Constructing instruction tables is usually described as "programming."[31]

Enlightenment probabilists like Condorcet conceived their mathematics of games as a reasonable calculus but not as one that could have been mechanically implemented by following rules without judgment or interpretation. Materialists like La Mettrie, inspired by automata, understood machinery in terms of clockwork gears, not as symbolic programs of instructions. The possibility of rigid rule following, obedient and undeviating, would have struck the Enlightenment probabilists as deeply unreasonable. Like Turing's human computer, the reasonable calculator must remember and affirm the rules, not just follow them blindly. To put the point paradoxically, within the framework of Enlightenment reason, calculation was reasonable but not rational: even for elementary arithmetical reckoning, rote rule following would not suffice. To put it less paradoxically but still (for today's readers) perplexingly, rules in general were not yet conceived as being first and foremost algorithms—a point that will be taken up in the next section.

Reasoning itself was imagined as a kind of combinatorial calculus, in turn understood as a form of deliberative intelligence, rather than as a substitute for it. For Leibniz, to calculate with the *characteristica universalis* would be willy-nilly to "reveal the reason in all things which was hitherto possible only in arithmetic."[32] In contrast to natural languages that permit ambiguity and false inferences, the artificial language of the *characteristica universalis* would enforce right reasoning on its users. "For people will be unable to speak or write about anything except what they understand, or if they try to do so, one of two things will happen: either the vanity of what they advance will be apparent to everybody, or they will learn by writing or speaking."[33] Calculation promoted rather than replaced understanding.

Even for (indeed, especially for) those Enlightenment thinkers most drawn to the possibilities of a mathematics of rational decision making modeled on probability theory applied to games, calculation by algorithms could serve as a model of intellectual clarity and even political autonomy—as in the case of Condorcet, whose probabilistic projects bear the closest resemblance to the ambitions of late-twentieth-century attempts to create a mathematics of rationality.[34] Condorcet's manuscripts contain many fragmentary plans for universal languages,[35] universal classification systems,[36] and even universal systems of legal contracts[37]—all based on the calculation of combinations and permutations. But Condorcet was quite capable of rejecting the results of calculation when they conflicted with "common reason" or seemed insufficiently grounded in observation.[38] Clarity must not be sacrificed to rigor, as he reprimanded an Italian political economist who had tried to quantify the desire to buy and sell.[39] More generally, Condorcet drew a distinction between mathematical calculation as a problem-solving tool and as a study "suitable for forming reason, for strengthening it."[40] Calculation was much more than a tool, much more even than a philosophical method for Condorcet. What calculation taught its practitioners was the "exactitude of the mind [*justesse d'esprit*]."

For Condorcet, exactitude of mind was an attainment that combined intellectual, moral, and political dimensions. In a textbook on arithmetic and geometry written for the public elementary schools he hoped that the revolutionary National Assembly would institute throughout the French republic, Condorcet used the simplest arithmetic identities—"three plus four equals seven"—to teach children the meaning of self-evidence and justified belief and thereby to liberate them from demagoguery and priestcraft: "From this, they will learn that the distinct memory of having had the perception of the identity of the two ideas that form a proposition, that is to say the self-evidence of this proposition, is the only motive they have to believe it . . . and that the memory of merely having always repeated or written this proposition, without having felt its self-evidence, is not a motive to believe."[41] In this fashion, simply by practicing the simplest arithmetic operations over and over again, children would learn about "the three intellectual operations of which our mind is capable: *the formation of ideas, judgment, reasoning.*" Reasonableness and autonomy went in hand-in-hand; rote learning was their common enemy. The numbers from one to ten must never be memorized, but instead taught "by intelligence and by reason; nothing is abandoned to routine."[42] Whenever these elements are manipulated in calculation, the mind must form anew a clear idea of their meaning as collections of units. In this way, Condorcet hoped, habit would not lead to mindless automatism.

Condorcet envisioned calculations applied to a vast range of problems, from contracts to tribunals to scientific hypothesis-testing. Like late-twentieth-century proponents of rational calculi, he believed that the mathematics of games might serve as the foundation for a quantitative science of the human realm. He was intoxicated by the possibilities of Leibniz's exhortation: "Let us calculate!" He is, in short, the most likely suspect for an advocate of reason as rationality *avant la lettre*, as Morgenstern, social choice theorist Duncan Black, economist Kenneth Arrow, and others recognized in their search for honorable intellectual ancestors. But even Condorcet did not understand reason as a set of formal rules—or rather, he did not understand the most formal of all formal rules, the algorithms of arithmetic, to be so formal as to be mindless. Nor did he conceive of the results of his reasonable calculus to be equivalent to reason itself, much less to trump reason: in cases of conflict, Condorcet (and other Enlightenment probabilists) preferred to modify the rules of their calculus rather than to challenge the verdict of conventional reason.[43] On this view, reason must consciously review, endorse, and criticize the rules that it obeyed—in contrast to material objects that conformed to natural laws without understanding them. Even for those Enlightenment thinkers who aspired to a calculus of reason, reason was distinct from automated, rule-bound rationality.

1.2. From Rationalization to Rationality

For Condorcet, even the algorithms of arithmetic were anything but mindless. As late as the 1890s, some authors on logic and mathematics still exhorted their readers to remember what the rules were about: "The algorithm knows only tokens [*Merkmale*]. . . . Therefore one must take care, especially these days, that one does not forget the essence of things in all that mathematics."[44] But the reminder to probe beneath the surface of the algorithm to its essence already rang hollow at the time these words were printed: by circa 1900, algorithms were understood as mechanical—in large part because machines could by then actually execute them. This came as a shock to those who still equated calculation with the exercise of the higher intellectual faculties. British mathematician Charles Babbage's contemporaries marveled over the implications of his difference engine, the most ambitious calculating machine heretofore contemplated:

> In other cases, mechanical devices have substituted machines for simpler tools or for bodily labour. . . . But the invention, to which I am adverting, comes in place of the mental exertion: it substitutes mechanical performance for an intellectual process: and that performance is effected with celerity and exactness unattainable in ordinary methods, even by incessant practice and undiverted attention.[45]

By the 1930s, however, Turing could propose a mechanical solution to German mathematician David Hilbert's *Entscheidungsproblem*[46] in which the calculator's state of mind was replaced by an instruction booklet:

We suppose, as in [argument] I, that the computation is carried out on a tape; but we avoid introducing the "state of mind" by considering a more physical and definite counterpart of it. It is always possible for a computer to break off his work, to go away and forget all about it. If he does this, he must leave a note of instructions (written in some standard form) explaining how the work is to be continued. This note is the counterpart of the "state of mind."[47]

But the history of ever more sophisticated mechanical computers (and the ambitious programs in artificial intelligence that they inspired[48]) is only half the story of how rationality became rule-bound. The other half concerns how rules themselves became increasingly identified with algorithms, and algorithms with mindlessness.

This development is surprising in light of the earlier history of rules. In both Romance and Germanic languages, the word for "rule" (French *règle*; German *Regel*; Italian *regola*; Dutch *regel*) stems from the Latin *regula* (from *regere*, to reign), which originally meant a rigid rod used to measure, compare, and correct lengths (compare with "ruler"). In each language, the literal and figurative meanings of the word ramified and blossomed from this root in diverse ways over centuries, but some generalizations hold across the board.[49] Under the influence of the monastic rule of Saint Benedict, laid down in the sixth century CE, the primary meaning of "rule" and its cognates throughout the Middle Ages and indeed well into the eighteenth century was a moral precept, a model or code of conduct for a way of life. The word *regula* (in the singular) of the *Regula Sancti Benedicti* applies to the order of monastic life as a whole, not to specific prescriptions contained in the individual chapters, which in any case are always subject to modification at the abbot's discretion.[50] A cluster of extended meanings relating to control, dominion, and government (all aiming at orderly conduct) accreted around this primary sense. By the fourteenth century, a strong secondary sense employed the word in the plural, as "rules" (*regulae*), to refer to a practical or procedural principle in an art or science: for example, canon law rules or grammatical rules or glaziers' rules or, especially interesting in light of subsequent developments, mathematical rules (e.g., the rule of three to find the fourth term of a proportion when three terms are given). The capacious category of rule could (and still can) embrace models, principles, laws, maxims, precepts, and guidelines, as well as algorithms. Its force was and is simultaneously descriptive (of an observed regularity) and prescriptive (of conduct to be regularly observed).

The "rule" (in the singular) was thus a model or pattern of conduct, instantiated but by no means exhausted by detailed instructions on how to eat, dress, sleep, work, and pray. This general sense of rule as model (often embodied by an exemplary person) persisted into the Enlightenment. As the article "*Règle, Modèle*" in the great *Encyclopédie* of Denis Diderot and Jean d'Alembert explained, "the life of Our Savior is the *rule* or the *model* for Christians: but . . . the counsels of the sages serve as a *rule* of our conduct: one would not say that they serve us as *model*, because it is properly speaking only a person, not actions, that serve as a *model*."[51]

The most controversial Enlightenment discussion about rules concerned the question of whether there existed rules for making and judging works of art and literature, analogous to logical or moral rules.[52] And perhaps the most influential of all contributions to this debate was Kant's discussion in his 1790 *Kritik der Urteilskraft* (*Critique of Judgment*), in which the distinction between "the rule" and "rules" served to demarcate the boundary between artistic genius on the one side and both scientific insight and artisanal skill on the other.[53] Genius in the fine arts is unrestricted by the rules that apply to mere talent or skill; on the contrary, it is genius "that gives the rule to art." The products of genius are "models [*Muster*], i.e., they must be exemplary; hence, though they do not themselves arise through imitation, still they must serve others for this, i.e., as a standard or rule [*Richtmaße oder Regel*] by which to judge." Since genius cannot explain its own workings, "it is rather as *nature* that it gives the rule." However admirable the works of the greatest scientific minds may be, they are not products of genius, according to Kant, because they can be learned by rules:

> For all of this could in fact have been done through learning as well, and hence lies in the natural path of an investigation and meditation by rules and does not differ in kind from what a diligent person can acquire by means of imitation. Thus one can indeed learn everything that Newton has set forth in his immortal work on the principles of natural philosophy, however great a mind was needed to make such discoveries; but one cannot learn to write inspired poetry, however elaborate all the precepts [*Vorschriften*] of this art may be, and however superb its models [*Muster*].

In this sense, science, even of the most exalted variety, is like mechanical skill: both can be mastered by diligence and "determinate rules"—whereas these are the necessary but not sufficient conditions for genius in the fine arts.[54]

However striking (and in the case of his judgment about Newton, strikingly odd) Kant's views about genius were, his nimble shifts between the singular sense of "the rule" as model, the inexplicable product of genius, and the plural sense of "rules," as explicit precepts to guide science and skill, parallels standard eighteenth-century usage. The singular sense of rule as model, an example to be imitated but not aped, still echoed the spirit of the monastic Rule of Saint Benedict. In contrast, the plural sense of "rules" pertained to detailed instructions—analogous to the procedures of carpenters or the regulations governing church benefices or the solving of an algebraic equation. Although Kant occasionally demeaned "technical rules" as "mechanical," he did not mean thereby that machines could execute them: "mechanical" did not yet mean "automatic."[55] In the seventeenth and eighteenth centuries, the word "mechanical" in English, French, and German still retained its associations with manual labor (as in Shakespeare's "rude mechanicals" in *Midsummer Night's Dream*).[56] In order to appreciate the force of the distinction between mechanical and automatic, it will be helpful to turn to the variety of Enlightenment rules that most closely approximates the definition that undergirds modern notions of rationality: the algorithm.

Until the mid-nineteenth century, the medieval Latin (originally derived from the Arabic) word *algorithmos*[57] and its cognates in other European languages referred exclusively to a special set of rules: the operations of arithmetic, addition, subtraction, multiplication, and division. From the mid-seventeenth century onward, there were notable, if not entirely successful, attempts to invent calculating machines to perform these operations.[58] But neither visions nor actual manufactured models of machines that in fact did calculate mindlessly led willy-nilly to widespread enthusiasm for reducing human reason or intelligence to algorithms in the first half of the nineteenth century. On the contrary, the immediate rationale for developing such machines was that if uneducated human laborers could be organized to do lengthy calculations, then calculation, *pace* Condorcet, was not an intelligent activity. This is how Babbage interpreted French engineer Gaspard de Prony's atelier for calculating logarithms according to the metric system instituted under the French Revolution.[59] The same manufacturing methods that Adam Smith had described for the pin factory (Prony's own inspiration) could be applied to calculation:

> We have seen, then, that the effect of the division of labour, both in mechanical and in mental operations, is, that it enables us to purchase and apply to each process precisely that quantity of skill and knowledge which is required for it: we avoid employing any part of the time of a man who can get eight or ten shillings a day by his skill in tempering needles, in turning a wheel,

which can be done for sixpence a day; and we equally avoid the loss arising from the employment of an accomplished mathematician in performing the lowest processes of arithmetic. [60]

The argument here is from economic rationalization: decompose the task into the simplest steps, divide the labor, hire the least skilled and cheap-

Figure 1.4. *Les Dames de la Carte du Ciel* (Women computers of the Paris Observatory), late nineteenth century. (Courtesy of Bibliothèque de l'Observatoire de Paris.)

est labor possible—and thereby increase efficiency while cutting costs. Far from there being a necessary connection between the actual mechanization of calculation and the conceptualization of reason as rule-governed rationality, the immediate impact of rationalization was to demote what had once been a scientific activity and even a definition of intelligence to the status of ill-paid, allegedly unskilled labor (which is why computational bureaus were so often staffed by women, from nineteenth-century observatories to World War II military projects [figure 1.4]).[61]

There were certainly protests against the devaluation of calculation from mindful to mindless exercise: makers of mathematical tables, actuaries, and other professional calculators invoked the "conscientiousness with which the work is performed" and the sage judgment required to make sense of the results.[62] But the long-term effect of deskilling calculation was to link economic rationalization with algorithmic rationality. Over and over again, in operations research and in linear programming (as we will see in chapter 2), in rational choice theory and in the observation of psychological experiments (as discussed in chapter 4), the pattern was repeated: first a complex task originally assigned to persons of seasoned experience and proven judgment was analyzed into its smallest component parts and sequenced into consecutive steps, then translated into simple instructions that could be executed by minimally trained workers whose discretion was tightly restricted, and finally taken over by a machine.[63] This was the historical arc that connected Babbage to Turing, allowing both to imagine an easy transition from a lowly worker who follows the rules to an exalted machine that could mimic "states of mind."

With the benefit of twenty-twenty hindsight, one can find in Babbage's works almost all the elements that over a century later were soldered together into Cold War rationality. He was an ardent promoter of the economic rationalization of both bodily and mental labor; he saw the far-reaching implications of algorithms for the execution of complex tasks; he even drew up the plans for machines, the difference and analytical engines, which could be programmed to perform astounding feats by means of algorithms. In his *Ninth Bridgewater Treatise* (1837), Babbage went so far as to draw an analogy between the functioning of his "calculating engine" and divine miracles: just as such an engine can be "ordered" by its maker to generate an unbroken series of square numbers over eons and at one crucial moment produce a cube without violating the "full expression of the law by which the machine acts," so the deity could have foreseen and ordained all apparent exceptions to natural laws from the moment of creation.[64] Even miracles did not violate natural laws. Yet it was neither the marvelous calculating engine nor the rules that governed it but rather the engine's maker that was the repository of rationality. For Babbage, the very fact that a machine could execute the algorithms disqualified them as expressions of higher intelligence or rationality.

1.3. Rules Rule

Babbage's example shows that it is a far cry from the economic rationalization of calculation to the creation of an ideal of rationality modeled on the rules of calculation. Why would a task assigned to poorly paid workers and ultimately to machines become the prototype for the loftiest flights of intellect? What changed, when, and why to make a rationality of algorithmic rules not just conceivable but irresistible? Depending on the variant of Cold War rationality that is taken as the terminus ad quem, different genealogies can be traced. One, stretching from George Boole to Alan Turing via Gottlob Frege, David Hilbert, Bertrand Russell, Alfred North Whitehead, and Kurt Gödel, connects efforts to secure logical foundations for mathematics and eventually to mechanize highly complex calculations and even proofs by means of computers;[65] another follows the proof of minimax theorems by Émile Borel and John von Neumann, culminating in Von Neumann and Morgenstern's *Theory of Games and Economic Behavior* (1944);[66] and still another emphasizes the stimulus given by military applications during World War II and the Cold War to applied mathematics and the formalization of strategy.[67] But there is the further question of how applications that had once struck even the mathematicians as far-fetched captured the imaginations of social scientists and military strategists. Game theory was slow to catch on among social scientists,[68] and many seasoned military men were downright hostile to formalist approaches to strategy.[69] How did the analogy between algorithmic rules and rationality become compelling?

Mathematical results alone did not suffice. Although Borel's work in probability theory in the 1920s and '30s anticipated important results for minimax solutions to two-person zero-sum games, Borel himself was notably skeptical about whether actual games—much less more complex social and economic situations analogized to games—would yield to such mathematical treatments.[70] Compare these reservations with the opening assertions of mathematical psychologist R. Duncan Luce and economist Howard Raiffa in their influential (probably more influential among social scientists than Von Neumann and Morgenstern's treatise[71]) *Games and Decisions: Introduction and Critical Survey* (1957):

We find today that conflict of interest, both among individuals and institutions, is one of the more dominant concerns of at least several of our academic departments: economics, sociology, political science, and other areas to a lesser degree. It is not difficult to characterize imprecisely the major aspects of the problem of interest conflict: An individual is in a situation from which one of several possible outcomes will result and with respect to which he has certain personal preferences. [72]

Luce and Raiffa acknowledged that such radical abstraction concerning the nature of rational behavior might take more empirically minded social scientists aback, but stood their ground:

One may object to treating this [economic situation] as a game on the grounds that the game model supposes that each producer makes one choice from a domain of possible choices, and that from these single choices the profits are determined. . . . However, in principle, it is possible to imagine that an executive foresees all possible contingencies and that he describes in detail the action to be taken in each case instead of meeting each problem as it arises. By "describe in detail" we mean that the further operation of the plant can be left in the hands of a clerk or a machine and that no further interference or clarification will be needed from the executive. [73]

"A clerk or a machine": what gave Luce and Raiffa the courage of their breathtaking convictions?

A rational reconstruction of the preconditions that might have made plausible the scenario Luce and Raiffa imagined would include the following: contingencies are exhaustively predictable; the course of action to be followed in case of each contingency can be optimized and reduced to a sequenced protocol of rules; the rules in question are algorithmic, in the sense that they can be executed without discretion or judgment, by "a clerk or a machine." For some of these preconditions, the history that made them possible (if not plausible) can be readily traced: for example, the rise of the algorithmic rule in logical and mathematical proofs;[74] or the application of the formalized subjective interpretation of probability theory to economic definitions of utility.[75] The salience of conflict situations in the wake of the most devastating war ever fought need not be belabored.[76]

More difficult to bridge is the gap between the complexity and contingency of not only social, political, and economic but also military situations and the calm assumption that these interactions could be adequately modeled, for example, by an n-person game in which all players are fully and equally informed about the game (what game theorists called "intelligence") and that optimal strategies for a given utility distribution can be achieved by strictly adhering to algorithmic rules ("rationality"). Game theorists acknowledged that their models were simplifications and might even lead to paradox, as in the case of the prisoner's dilemma (as we will see in chapter 5), but did not abandon ship: "Of course, it is slightly uncomfortable that two so-called irrational players will both fare much better than two so-called rational ones [in the prisoner's dilemma]. . . . No, there appears to be no way around this dilemma. We do not believe that there is anything irrational or perverse about the choice of α_2 and β_2 [i.e., both prisoners decide to defect] and we must admit that if we were actually in this position we would make these choices."[77] Rationality as rule following had become plausible both as prescription and description.

Were there developments internal to the mid-twentieth-century social sciences that would have fortified the analogy between the rationality of rule following and human conduct in circumstances of uncertainty? More precisely, what was the status of rules as objects of social scientific inquiry prior to the advent of game theory and rational choice theory in the mid-1950s? It may not be coincidental to the prominence of rules in postwar American sociology that New Deal legislation had triggered the greatest expansion of government rulemaking in American history, a trend that continued unabated into the 1970s.[78] But the private sector was also increasingly regarded as a congeries of rules. Amongst American sociologists and political theorists, the analysis of rules flourished in the study of bureaucracy, perhaps most influentially in Alvin Gouldner's *Patterns of Industrial Bureaucracy* (1954), which described bureaucracies in terms of the kinds and functions of rules, based on a case study of a gypsum mine and factory.[79] As one of the book's reviewers observed, the book centered around the theme of how community customs in the mine hardened into bureaucratic rules in the factory: "His study is highly illuminating in showing the bases upon which human behavior become regularized and routinized in the factory."[80] Directly following Gouldner's essay in a 1952 *Reader on Bureaucracy* came Herbert Simon's call for a complete overhaul of the principles of administration in order to turn it into a genuine science. Empirical studies modeled on F. W. Taylor's experiments in increasing industrial productivity and theoretical specifications on the limits to rationality, defined as the maximization of efficiency, were urgently required. For every administrative situation, Simon asserted that there was a unique rational decision under the given constraints: "Two persons, given the same skills, the same objectives and values, the same knowledge and information, can rationally decide only on the same course of action."[81]

Gouldner's and Simon's brief responses drew the axis around which subsequent studies of bureaucracy in postwar American sociology would revolve. The transition from informal customs to formal rules on the one hand, and the tension between rigid rules and discretion on the other became dominant themes in a vast literature on modern bureaucracies.[82] (The same tension, framed as clinical judgment versus actuarial formulas, greatly exercised psychologists and psychiatrists, starting in 1954 with Paul Meehl's *Clinical versus Statistical Prediction*.[83]) Rules remained central to the analysis of bureaucracies, but their primary characteristic shifted from impersonality in the name of impartiality (as in the Weberian theory of bureaucracy) to formality in the name of rationality—even in those social sciences that did not quickly succumb to the charms of more mathematical models of rational interaction.

In the 1950s and '60s, the most unlikely objects—common sense, culture, even irrationality—were redescribed as ensembles of rules by social scientists, including critics who doubted that game theory or rational choice theory or other formal models of rationality could be applied to real world situations (figure 1.5). Sociologist Harold Garfinkel, although skeptical about the possibility of reconciling what he called "scientific" and "common sense" rationalities, affirmed that "the definition of credible knowledge, scientific or otherwise, consists of rules that govern the use of propositions as further grounds of inference or action";[84] psychologist Robert Bales (whom we will encounter again in chapter 4) described "common

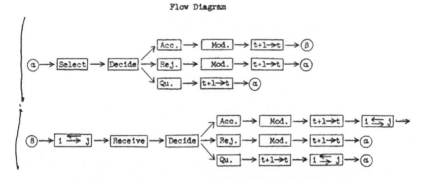

Figure 1.5. Flow diagram of a two-person decision-making process independent of its content. (Robert F. Bales, Merrill M. Flood, and Alston S. Householder, "Some Group Interaction Models," RAND RM-953 [1952], 15. (Reproduced with permission of RAND Corporation.)

culture" as "the set of rules, or programs, or norms, more or less common to all [group] members."[85] Political scientist Sidney Verba, albeit insistent on the limitations of models of rationality to describe the international political situation, was pessimistic about the utility of clinical studies of nonrationality to help correct these deficiencies. Such studies were unlikely to "supply us with rules . . . to enable us to predict what particular non-logical force is likely to be more widespread among what groups."[86] Rules ruled everywhere.

1.4. The Splintered Self

The debate over Cold War rationality careened between meticulous calcula-tion and wild guesswork. It was precisely within this roller-coaster world of axioms and approximations that the rationality of rules cast its soothing spell. Although figures like Kahn (see chapter 3) could trumpet a terrifying self-confidence in their speculations about nuclear war,[87] most of the Cold War rationalists were acutely aware of the limitations of their models. In a 1971 memorandum to the United States Arms Control and Disarmament Agency, Morgenstern for example pointed out that errors in the ABM (an-tiballistic missile) system were inevitable, no matter how sophisticated the models of command and control: "It is time for the government to wake up to the fact that we live in a stochastic universe, that error is a perma-nent and inherent feature of the world."[88] Several close calls had set off NORAD (North American Aerospace Defense Command) alarms that the United States was under attack, so Morgenstern's warning was more than just a theoretical possibility: "On October 5, 1960, the warning system at NORAD was under massive attack by Soviet missiles with a certainty of 99.9 percent. It turned out that the BMEWS [ballistic missile early warning system] radar in Thule, Greenland, had spotted the rising moon. Nobody had thought about the moon when specifying how the system should act."[89] An official RAND brochure commemorating the corporation's first fifteen years of operation struck a note of "modesty and realism" with re-spect to what its analyses could accomplish, given that "objective elements" quickly became "too complex to handle" and many "subjective elements" escaped formal models altogether. Yet "to abandon our efforts at rational analysis, unsatisfactory as these are, would merely uncover the uncomfort-able truth that the alternatives also entail analysis—and analysis which may be far more difficult." The alternatives, which included "reliance on authority" and "intuitive judgments based on experience," would simply trade explicit, rational analysis for the implicit, irrational sort, with no gain in accuracy.[90] The Cold War rationalists rarely forgot the limitations of their rules and tools, however Panglossian their optimism about future improvements and applications waxed. But they also insisted that their methods, however abstract and approximate and oversimplified, were better than the alternatives.

Some, like Flood, hoped that experimental game situations would both test and improve the skeletal and (for anything more complicated than a two-person zero-sum game) inconclusive machinery of game theory.[91] The RAND Corporation pioneered a minor industry of simulations, ranging from war games played on boards to group interactions observed through one-way mirrors.[92] Economist Thomas Schelling, who had served as a foreign policy advisor to the White House and negotiator with European nations over NATO provisions in the early 1950s before taking a faculty position at Yale in 1953, was particularly alert to the importance of enriching formal rationality with considerations derived from hands-on experience of negotiation, bargaining, and strategy. He underscored the inevitability of social effects. Postulating the coolest rationality on all sides of a conflict would not eliminate the effects of what he called "social perception," even under conditions of isolation, anonymity, and "absolute silence," whenever more than one mind was involved: "An analyst can deduce the decisions of a single rational mind if he knows the criteria that govern decisions; but he cannot infer by purely formal analysis what can pass between two centers of consciousness."[93] Schelling nonetheless held fast to the hope that such experiments would salvage rather than sink formal approaches to rationality.

There was no end of critics who disagreed, on technical, philosophical, or moral grounds.[94] But as we will see in the following chapters, the Cold War rationalists were in some ways their own severest critics, uneasily aware of the vast reality that their rule-governed rationality could not cover or even contradicted. They did not fear individual subjectivity, so long as it took the form of paired preferences that obeyed conditions of consistency. Individual tastes, no matter how eccentric, and individual values, no matter how extravagant, were granted free rein—if they preserved transitivity and preferences dictated choices.[95] Their greatest challenge was to bridge individual and collective rationality. To their disappointment, collective preferences proved almost impossible to integrate, breeding paradoxes, dilemmas, and impossibility theorems. Individually rational voters could, for example, become flagrantly irrational as a group—not because they succumbed to mob psychology but because their preferences violated transitivity. Economist Kenneth Arrow stated the problem with lapidary concision: "If consumers' values can be represented by a wide range of individual orderings, the doctrine of voters' sovereignty is incompatible with that of collective rationality." Only an individual with dictatorial powers could preserve what Arrow called "collective rationality."[96]

But even a dictator might splinter into multiple selves: neurotic, capricious, or just weak-willed. Schelling recognized the catastrophic implications of a nonintegrated self for situations of conflict, bargaining, and deterrence. "If we confine our study to the theory of strategy, we seriously restrict ourselves by the assumption of rational behavior—not just of intelligent behavior, but behavior motivated by a conscious calculation of advantages, a calculation that in turn is based on an explicit and internally consistent value system."[97] The irrationality most damaging to Cold War rationality was not the passions or malevolence or caprice; it was inconsistency, two (or more) souls within a single breast competing for mastery to choose and decide. Schelling hoped that it would turn out that not only criminals, neurotics, and Communists, but even his own fractious children would be revealed to be rational in their irrationality, and analogously so: "It may be easier to articulate the peculiar difficulty of constraining a Mossadeq by the use of threats when one is fresh from a vain attempt at using threats to keep a small child from hurting a dog or a small dog from hurting a child."[98]

Cold War rationality breathed new life into the ancient moral conundrum of *akrasia*, already articulated by Plato in the *Protagoras*, of a will too weak to act for its own good.[99] The new examples bore the stamp of their times: no longer Odysseus binding himself to the mast to fortify his will against the temptations of the sirens but a gambler trying to stay out of casinos—or the United States government deciding to get out of Vietnam.[100] Whether played out in microcosm or macrocosm, the problem was perceived as the same: how to forge the internal consistency of society and self that would make the world safe for the rationality of rules.

Rationality and rationalization had chimed with one another since the industrialization and later mechanization of calculation in the nineteenth and twentieth centuries. The next chapter examines how the "rules of action" developed to streamline and economize military operations during World War II became the backbone of a new applied science, operations research, and the inspiration for the symbolic processing rules of linear programming. As in the case of the human computers described by Babbage and Turing, standardized routines performed by low-level laborers became the template for high-level rationality.

TWO

The Bounded Rationality of Cold War Operations Research

Frankfurt/Rhein–Main US Air Force base 1948: On his seventieth day on the LeMay Coal and Feed Run, Lieutenant Fred V. McAfee, a self-described flour and dehydrated potatoes man who "was not above hauling macaroni," iterated his very regular life:

> They briefed me on the basic pattern to follow and I've done it so many times since, I could repeat their lecture. I was assigned a 6,500-foot altitude. I had ten tons of flour, used 2700 rpm and forty-five inches of manifold until I broke ground and then I throttled back to 2500 rpm and forty inches. At 500 feet I dropped her back again, this time to 2300 rpm and thirty-four inches. I climbed on a course of 180 degrees for seven minutes, at an indicated speed of 160 miles an hour, leveled off at my assigned altitude, where my power settings were reduced still again to 2050 rpm and twenty-nine inches. My indicated air speed was 170 miles an hour. I'd picked up Darmstadt beacon. Presently it was Aschaffenburg, then Fulda's range and presently I was in the corridor [see figure 2.1 for the radio beacons that guided planes to the air corridors]. I let down over Tempelhof at 170 miles an hour, pushed her in and watched them unload. Then back here. Then Berlin. And so it's been for seventy days. A very regular life.[1]

The regularity of McAfee's life was a product of an assembly-line style of scientific management that the US Air Force used in Operation Vittles to get essential supplies to the blockaded zones of western Berlin in 1948 and 1949. The management methods used at the USAF air bases for the logistics of the airbridge harked back to the early twentieth-century time studies of Frederick Winslow Taylor and motion studies of Frank and Lillian Gilbreth. At the air force headquarters at the Pentagon, however, a group of applied

HOW THE BERLIN AIR LIFT WORKS

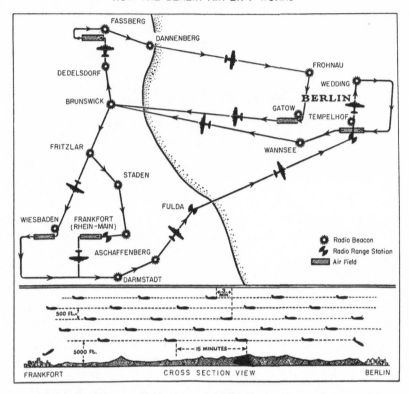

Figure 2.1. Plan for the Berlin airlift highlighting the British and American bases from which the planes took off, the radio beacons in the western zones, the boundary line with the Russian zone of eastern Germany, the three air corridors through that zone, and the Gatow and Tempelhof airports in the British and American sectors of Berlin, respectively. Flying at a designated altitude enabled planes to land every three minutes at Tempelhof. (*New York Times*, Paris Bureau, US National Archives, RG 342-G, Box 25.)

mathematicians in the Project for the Scientific Computation of Optimum Programs (Project SCOOP) was using the Berlin Airlift to create a new science of managing. The mandate of Project SCOOP was to mechanize the planning process by constructing computable algorithms that could determine the *best* time-staged deployment of personnel and materials for a military operation such as the Berlin Airlift. With the aid of mathematics and electronic digital computers, the Project SCOOP team hoped to achieve through centralized planning the optimal combination of resource items and production activities that a decentralized, competitive market structure would have achieved. The ingenious design of an algorithm for deriving economically rational decisions from an equation system that included an objective function maximizing, for example, the tonnage delivered to Berlin and a matrix quantifying the interdependency of inputs and outputs was, however, years ahead of the electronic digital computing capacity necessary to implement the algorithm. As we will see, with their punch-card calculators in 1948 the only *optimal* program that Project SCOOP could determine was for the least cost diet that could provide the essential nutritional needs of an active, urban economist weighing seventy kilograms; Operation Vittles had to be planned with suboptimal protocols.

Scarce computing capacity limiting the scope of the optimizing mathematics became a dominant trope in subsequent Cold War US military-funded research, including the production planning for Project SCOOP and Office of Naval Research done by Herbert Simon and his economist colleagues at the Carnegie Institute of Technology in the 1950s. The Carnegie research team had to adapt their maximizing models so that the solutions could be obtained with existing, limited computational resources. Intrigued by how that practical limitation shaped theoretical developments, Simon conceived of a framework of "bounded rationality" and called for the development of descriptive models of rational behavior based on "satisficing" (making do with reasonable aspiration levels) rather than "maximizing." For Simon, the economists' narrow focus on rationality as a quality of the choice outcome had to be broadened to include the "procedural rationality" that was a quality of the decision process and sensitive to the costs of searching and computing. Thus, while holding out the promise of an optimal allocation of resources in the absence of any market, the mathematical programming protocols that Project SCOOP initiated also nurtured Nobel Prize–winning musings on bounded rationality. But first back to Berlin.

2.1. Operation Vittles

The shattered, occupied, and zoned city of Berlin was a major—if not the key—battleground for the Cold War. There were no hot war battles between the two superpowers in Berlin, but it was a significant setting for their brinkmanship, crises, and settlements. In 1944, a year before the end of World War II, allied powers in the European Advisory Commission determined the postwar zones in Germany and Berlin that would be occupied by the Soviet Union, the United States of America, and Britain. The Yalta conference in 1945 accorded France sectors carved out from the planned US and British jurisdictions. Berlin was surrounded by the land intended for Soviet occupation, but the Commission established the traffic routes by which American, British, and French garrisons in the respective Berlin sectors could be supplied: a twenty-mile-wide air space for each of three air corridors, a railway line from Helmstedt, and a highway from Marienborn. After Germany's surrender on May 8, 1945, the zoned occupation proceeded according to plan.

In the spring of 1948, the American, British, and French occupying forces took steps toward a more unified and autonomous Western Germany, including issuing new currency for their combined zones in Western Germany. The Soviet Union, worried about the potential for a separate, powerful German state, responded with increased restrictions on travel to Berlin and insisted that the currency of the Soviet zone be the sole currency for Berlin. On June 23, the Western allies announced their plan for a new *deutsche mark* in the French-, British-, and American-occupied zones of Berlin. That same day, the Soviet authorities issued the East German mark (also called the *deutsche mark*, colloquially referred to as the *ostmark*) with the intention of it being the currency for all of Berlin. The Soviet occupying forces began a complete blockade of road, rail, and river traffic to and from the American, British, and French sectors of Berlin and cut off key sources of electrical power to those sectors.

Britain, France, and the United States were faced with the choice of withdrawing from Berlin or remaining at a high cost. The US government decided "to remain in Berlin, to utilize to the utmost the present propaganda advantage of our position to supply the city by air and, if the restrictions continued, to protest to the Soviets and keep the Berlin situation before World attention."[2] On Sunday, June 26, 1948, the Berlin airbridge (*Luftbrücke*) began with the USAF Operation Vittles and the British Operation Plane Fare. The American and British airlifts were ultimately successful in getting sufficient supplies to the western zones of Berlin for the two and a half-million people who lived there.[3] On May 12, 1949, the Soviet Union lifted the road blockade in exchange for an agreement to a meeting of the respective four foreign ministers. In order to build up a cushion for possible future stoppages, Operation Vittles continued until the last official flight on September 30, 1949. For most of that time, Major General William H. Tunner, initially serving under General Curtis LeMay, the commander of the US Air Force in Europe, was the commanding officer for Operation Vittles.[4]

Tunner and his staff measured, evaluated, and dictated the procedures for loading, flying, and unloading aircraft. They used physical models of the layouts of the airspace altitudes and airfield facilities to study bottle-

Figure 2.2. The plotting board of the airlift terminal at the British base at Fassberg, Germany, "helps loading officers to study new methods of operations before putting them to work on a large scale. The models, which may be moved to test loading plans, include US Air force C-54s, British trucks, and German trains and track sidings. Although an RAF base, all Berlin bound cargo is flown from Fassberg in US Air Force planes." (US Air Force, US National Archives, RG 342- G, Box 25.)

necks and simulate improved operations (figure 2.2). Tunner's philosophy was that achieving a "precise rhythmical cadence . . . constant as the jungle drums" determined the success of an airlift; after analyzing the data he insisted on only a three-minute interval between takeoffs because it provided "the ideal cadence of operations with the control equipment available."[5]

Figure 2.3. Unloading planes at Tempelhof, October 28, 1948.
(*New York Times*, Paris Bureau, National Archives, RG 306-PS, Box 85.)

The boring regularity of Lieutenant McAfee's piloting days on the LeMay Coal and Feed Run was exactly as planned (figure 2.3).

Before Tunner had taken over in late July 1948, the air force had assigned minimum amounts of cargo to be delivered. On October 14, 1948, the day before he left to assume his iconic Cold War role as chief of the Strategic Air Command, General LeMay signed an agreement with the British Air Forces of Occupation for a merger of their operations under Tunner's command with the primary mission being the delivery to Berlin, "in a safe and efficient manner, the maximum tonnage possible, consistent with combined resources of equipment and personnel available."[6]

The switch from the goal of minimum quotas by aircraft to maximum tonnage by operation, the precisely managed cadence, Tunner's leveraging of data analysis, and his encouragement of tonnage competitions between squadrons (figure 2.4) ensured that Operation Vittles brought 2.3 million tons of food to West Berlin via 277,500 flights.

At the peak of the airlift, planes were landing in the Western zones of Berlin at a rate of one every sixty-two seconds. Tunner's scientific management, however, was limited to the careful analysis and improvement of

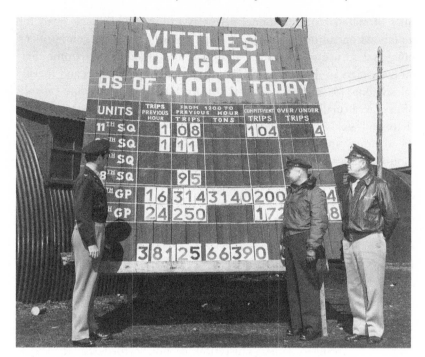

Figure 2.4. The caption of the undated USAF photograph reads:
"Fassberg RAF Station, Germany. Theron C. Coulter, left C.O. 60th Troop Carrier Wing
and Lt. Col. Conway S. Hall, C.O. 313th Troop Carrier Group, 2nd from left, viewing the
'HOWGOZIT' board which keeps Fassberg personnel informed from hour to hour as to the
number of flights and tonnage flown to Berlin." (US National Archives, RG 342-G, Box 25.)

single-purpose operations with given personnel and equipment. The maximums that Tunner strove for were similar to what mathematicians would describe as "local" ones, and his staff's capacity for weighing alternatives was extremely limited.

The high cost of Operation Vittles (accounting for 14 percent of the entire USAF budget) came at a time when the US government was under considerable pressure to cut military budgets and steer resources to a still recovering peacetime economy. Operation Vittles validated the air force comptroller's plan initiated in 1947 to increase the cost-effectiveness of air force operations through the mechanization of the planning process. The US military had long-standing procedures for what they referred to as "programming" military operations.[7] Planners in the Pentagon headquarters used rules of thumb, judgment based on experience, and arbitrary decisions to construct a time-phased schedule of activities and of quantities of material and personnel necessary to meet the strategic goals for a planned operation. The new approach envisioned by the comptroller's office was to use mathematical protocols and eventually electronic digital computers to determine the *best* combination of activities and logistical schedule for an operation and construct a flexible procedure capable of easy recalculation of a new optimal schedule of action in response to changes in goals or resources. With this research, management science in the air force, and subsequently, in US industry, went from the time and motion studies for improving singular output with given inputs to the computer-enhanced optimal allocation of alternative inputs to alternative outputs.[8]

2.2. Project SCOOP

During World War II, Captain Charles "Tex" Thornton created the US Army Air Force Statistical Control unit to improve budget control and planning by using statistical information for management decisions and operations control. The Statistical Control group employed, among others, George Dantzig,[9] Robert McNamara, and Harvard business professor Edmund Learned. According to Dantzig's historical account, Learned developed what the group called "a program for programming"—a generic time-phased schema that would connect the scheduling of many command agencies in any detailed air force operation (such as that pictured in figure 2.5).[10] In Learned's World War II framework, improvements in reducing the time it took to develop a new program for a major military operation emerged from the sequential ordering of some forty-six major steps for determining planning factors on requirements for and attrition and consumption rates of equipment and personnel. Learned's schema for the echelon-to-echelon unidirectional flows of information on planning factors mirrored the bureaucratic structure of the USAF.

Despite the care the Statistical Control group took in constructing the schema, they estimated that after World War II it would take seven months to complete the process of detailed programming for a major new operation—too long in the nuclear era. Also, Learned's framework could not solve the *economic* problem of planning because it did not take into consideration *alternative* uses of resources to determine the most cost-effective way to achieve an operational goal.

Under the 1947 plan to separate the army air forces from the army and constitute it as an autonomous branch of military service, the office of the

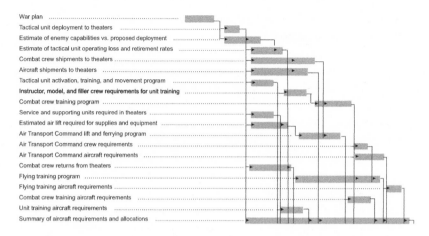

War plan

Tactical unit deployment to theaters

Estimate of enemy capabilities vs. proposed deployment

Estimate of tactical unit operating loss and retirement rates

Combat crew shipments to theaters

Aircraft shipments to theaters

Tactical unit activation, training, and movement program

Instructor, model, and filler crew requirements for unit training

Combat crew training program

Service and supporting units required in theaters

Estimated air lift required for supplies and equipment

Air Transport Command lift and ferrying program

Air Transport Command crew requirements

Air Transport Command aircraft requirements

Combat crew returns from theaters

Flying training program

Flying training aircraft requirements

Combat crew training aircraft requirements

Unit training aircraft requirements

Summary of aircraft requirements and allocations

Figure 2.5. Portion of schematic diagram showing twenty of the forty-six major steps in Air Force wartime program scheduling before Project SCOOP. (Marshall K. Wood and Murray A. Geisler, "Development of Dynamic Models for Program Planning," *Activity Analysis of Production and Allocation: Proceedings of a Conference,* ed. Tjalling C. Koopmans [New York: John Wiley & Sons, 1951], 191.)

Deputy Chief of Staff/Comptroller General E. W. Rawlings took charge of program planning, statistical control, and budgetary functions. In June 1947, Rawlings formed a group in the Planning Research Division of the Directorate of Management Analysis at the Pentagon to mechanize program planning for the air force by developing decision-making protocols that could realize the future potential of electronic digital computers. That initial group included George Dantzig (chief mathematician), Marshall Wood (chief economist and head of the Planning Research Division), and Murray Geisler (head of the division's Standard Evaluation Branch); others joined the effort soon after. On October 13, 1948, General Hoyt Vandenberg, the USAF chief of staff, conveyed to the entire air force the order of the secretary of the air force that defined the "scope and character" of the newly named Project for the Scientific Computation of Optimum Programs:

a. The primary objective of Project SCOOP is the development of an advanced design for an integrated and comprehensive system for the planning and control of all Air Force activities.

b. The recent development of high speed digital electronic computers presages an extensive application of mathematics to large-scale management of problems of the quantitative type. Project SCOOP is designed to prepare the Air Force to take maximum advantage of these developments.

c. The basic principle of SCOOP is the simulation of Air Force operations by large sets of simultaneous equations. These systems of equations are designated as "mathematical models" of operations. To develop these models it will be necessary to determine in advance the structure of the relationships between each activity and every other activity. It will also be necessary to specify quantitatively the coefficients which enter into all of these relationships. For this purpose the structure will need to be analyzed and the factors evaluated with much greater precision and thoroughness than has ever before been attempted in Air Force program planning.

d. Important advantages for the Air Force may be anticipated from SCOOP: First, the procedures to be developed will free the staff from its present preoccupation with arithmetic. Second, since the implications of alternative courses of action may be worked out in detail, it will permit the staff to concentrate attention on the appraisal of alternative objectives and policies. Third, attention will also be focused on factors: a premium will be put on the development of more efficient operating ratios. Fourth, the intensive analysis of structure, which is a prerequisite for the new procedures, will permit the integrated and parallel development of programs and progress (or statistical control) reports; this will permit a more rigorous control of actual operations.

e. The objectives of SCOOP cannot be achieved immediately, but are to be attained gradually over a period of years.[11]

The air force directive ordering all echelons of the air force to support the new project for achieving optimum programs was announced a day before LeMay signed the agreement with the Royal Air Force to put both airlift operations under Tunner's command with a goal of maximizing tonnage delivered.[12] Project SCOOP's scientific computation offered the prospect of a more cost-effective, dynamic approach to maximization than had been achieved before. The Pentagon-based project initiated a new applied mathematics of *optimization*—the decision-making process that determined with mathematical models and computable algorithms the way in which alternative inputs and activities could be combined to achieve a goal of maximum output (or minimum cost), subject to constraints. There are two important features to highlight here: from the beginning, the design of optimum programs for the air force was under the auspices of the comptroller's office, and was thus a part of the budgetary planning and management analysis branch of USAF headquarters at the Pentagon; and the modeling and solution strategies were designed for electronic digital computers that would not be available for air force programming until the early 1950s. The context for the first feature was President Truman's insistence that the military budget had to be cut; there was considerable pent-up demand for a thriving consumer-based economy, the electorate would not tolerate increases in military spending for a possible future war, and the US monopoly on nuclear weapons ostensibly guaranteed national security.[13] The air force, along with other military branches, had to do more with less. Project SCOOP aimed to combine the science of economizing with effective computation.

2.3. Dantzig's Linear Programming Model and Simplex Algorithm

The kernel of Project SCOOP's plan to mechanize decisions for determining the best schedule of action for military operations was George Dantzig's formulation in June 1947 of a linear programming model, his construction in August 1947 of the simplex algorithm for solving linear programming problems, and the subsequent coding of the simplex algorithm for digital computers.[14] The National Applied Mathematics Laboratories of the National Bureau of Standards (NBS) assisted the Pentagon team with the latter task.

Dantzig's model, which he described as a "linear technology," consisted of a linear "objective function" that stated mathematically that the sum of the outputs of the activities over a specified time period was to be at its maximum (or total costs were to be at their minimum); linear constraints in the form of equations that specified the technological relations between resource items that served as inputs to and outputs from production activities; and linear constraints in the form of inequalities that specified, for example, maximum available resource limits for inputs.

The objective of the linear programming model, to maximize gain or minimize loss through the best allocation of given resources, was consistent with an economist's representation of rationality. The modeling of the interdependencies of the components of the system was also heavily influenced by economics and in particular by the Nobel Prize–winning research of the Russian-born economist Wassily Leontief. While a professor of economics at Harvard University in the 1930s, Leontief had constructed quantitative input/output tables for the US economy that, for example, accounted for the fact that a given quantity of steel production required a given quantity of coal production. Concerned about repercussions from massive demobilization that would likely occur at the end of the war, the US Bureau of Labor Statistics had hired Leontief in 1941 to begin construction of a large interindustry input/output model of the US economy to measure the likely effects of demobilization on employment in different industries.[15] Dantzig and his colleagues at Project SCOOP appropriated Leontief's matrix framework by conceiving of the air force as "comprising a number of distinct activities. An Air Force program then consists of a schedule giving the magnitudes or levels of each of these activities for each of a number of time periods, such as weeks, or months, or quarters within the larger general time interval covered by the program."[16] Each activity required and produced *items* such as trained personnel or equipment. Each distinct production activity was a column in the matrix and each item was a row. Data collected by the comptroller's office would provide the coefficients in the matrix cells indicating, for example, how many aircraft were needed to fly ten thousand flights in three months from Fassberg to Berlin.

Given the economists' assumption that *Homo economicus* maximizes gain or minimizes loss and the relevance of Leontief's interindustry model of an economic system, Dantzig also hoped that economists could supply him with an algorithm for solving military programming problems, which were essentially problems in the efficient and optimal allocation of scarce resources for a system. There was a decades-long tradition of using mathematics in descriptive economics to abstractly demonstrate that maximizing self-interest could lead to an efficient allocation of resources and an optimal point of tangency connecting consumers' and producers' interests. In June 1947 Dantzig met with the Dutch economist Tjalling Koopmans at the Cowles Commission at the University of Chicago.[17] As a statistician for the British Merchant Shipping Mission in Washington during World War II, Koopmans had worked on a programming problem of minimizing the number of ships needed to deliver a fixed amount of supplies for the war effort. Dantzig learned on his visit to the Cowles Commission, however, that Koopmans and others working on "normative," prescriptive, system-wide allocation problems had not come up with an efficient way of computing numerical optimal solutions to complex cost-minimization or output-maximization problems. Noncomputability had forced Koopmans and other economists to abandon or resort to approximating trial-and-error solutions in their pursuit of a mathematical means for *planning* an optimal allocation of resources in a system.

In August 1947 Dantzig slew the dragon of noncomputability with his iterative "simplex" algorithm.[18] His algorithm had all the qualities of the ideal algorithm that, as we saw in chapter 1, A. A. Markov had eloquently praised: prescriptive precision, generality, and orientation to a desired result. For the first time, operations and economic researchers working on systematic optimal allocation problems were able to echo Gottfried Leibniz in saying, "Let us calculate, Sir." This new calculation-for-allocation capacity would go far in putting the "science" into management science and economic science in the 1950s and 1960s. In his August 5, 1948, briefing to the air staff, Dantzig asserted:

> One ranking mathematical economist at a recent conference at Rand [*sic*] confessed to me that it had remained for Air Force technicians working on the Air Force programming problems to solve one of the most fundamental problems of economics. What he meant is that the techniques which we are developing are equally applicable to planning in any large organizational structure, i.e., the Air Force, other Military Establishments, the National Economy (for Industrial Mobilization) and Potential Enemy Economies (for finding best means of the their neutralization).[19]

As Koopmans acknowledged in his Nobel Prize autobiography and his 1975 prize lecture, the initial conversation with Dantzig in the summer of 1947 and the contacts that were soon to follow proved very fruitful for Koopmans, the Cowles Commission, and the economics discipline.[20] Dantzig's successful articulation of what appeared to be an efficient algorithm for solving optimization problems as well as his demonstration of the pressing air force interest in the science of economizing led Koopmans to broaden his shipping problem into a generalized "activity analysis" and led the Cowles Commission, now under Koopmans's direction, to seek a major grant from the US Air Force via the RAND Corporation for research into the "theory of resource allocation." The new focus on "optimal behavior" first appeared in the Cowles Commission annual report for 1948 and 1949 and stands in stark contrast with the foci in the 1947 report. The 1948–1949 report describes the study of optimal economic behavior also known as "welfare economics" as "a normative science: it starts with the accurate formulation of some objective to be regarded as economically good for society and derives rules of behavior from that objective."[21]

During the exploratory months of 1947, Dantzig also learned of the close connection between his linear programming approach and the game-theoretic approach that the mathematician John von Neumann and the economist Oskar Morgenstern had introduced in their *Theory of Games and Economic Behavior* (1944). Both analytical frameworks were directed toward explicit, optimal decision making via the quantitative evaluation of alternative outcomes. In his first meeting with Dantzig on October 3, 1947, John von Neumann speculated that game theory and Dantzig's linear programming were analogues of each other.[22] Within two years of meeting Von Neumann, Dantzig and other mathematicians had proved that every two-person zero-sum game could be turned into and solved as a linear programming problem.[23] As we will see in chapter 5, outside of that limited class of games, optimal solutions could be as elusive as they were for computationally-strapped mathematical programming problems.

2.4. Project SCOOP's Limited Computational Capacity

As with its Cold War fraternal twin, game theory, which we will encounter in chapter 5, linear programming was based on the mathematics of convex sets, and the simplex algorithm for air force programming depended on the manipulation of often large matrices with numerous multiplications. Multiplication speed was a key limiting factor on practical computation, and stored-program electronic digital computers were a necessity for dramatically improving multiplication speed.[24] The SCOOP team was heavily involved in examining the engineering alternatives for machines to reduce multiplication time. Dantzig tested experimental circuits and in October 1948 began planning with the National Bureau of Standards (NBS) for an expected 1951 delivery of the first UNIVAC computer designed by J. Presper Eckert and John W. Mauchly.[25] In 1948, the air force comptroller, General Rawlings, awarded the NBS $400,000 ($3.8 million in 2012 dollars) to design a quickly constructed interim computer before the UNIVAC was ready. The NBS's SEAC was the first electronic computer to solve a small linear programming problem, but that did not occur until January 1952 and the limitations of the SEAC were such that it could not be used for programming military operations.

Until 1952, the Project SCOOP team was only able to compute a truly optimal solution to one nontrivial linear programming problem: In early 1948, the NBS staff used IBM electromechanical punch-card calculators to compute what would have been the most economical diet for an active man in 1939. In 1941 in an unpublished Bureau of Labor Statistics memorandum, Jerome Cornfield had tried to formulate a linear program to find the least costly diet that would meet the nutritional needs of a typical soldier, but he did not have a computationally efficient algorithm for solving for the optimum. Dantzig, a friend of Cornfield, took up this challenge using data from George Stigler's 1945 attempt to find the minimum-cost daily diet combination of 77 different foods that met the recommended daily nutritional requirements of nine nutrients for a 154-pound economist living in a large city for the years 1939 and 1944. Stigler stated his cost minimization problem in the form of nine equations in 77 unknowns. Stigler also did not have computational capacity for an optimal solution so he had to make do with a "clever heuristic," as Dantzig described it, to approximate a solution. When the NBS staff solved Stigler's diet problem with Dantzig's simplex algorithm, calculators, and many staff-hours, they determined that the optimal diet in 1939 was $39.69 a year versus Stigler's estimate of $39.93.[26]

Although this computation was not directly relevant to air force operations, it was essential to the viability of Project SCOOP because it was the first major test of the computational efficiency of Dantzig's simplex algorithm. Searching for and testing for the most efficient way of computing a solution engendered its own analysis of optimal *procedures* and its own rationalization of algorithmic design and the production process for calculation. With echoes of Gaspard de Prony's atelier for calculating logarithms, which we encountered in chapter 1, the feat of finding the minimum cost of a diet supplying essential nutrients required nine statistical clerks working the equivalent of 120 staff days to perform the required 17,000 multiplications and divisions using desk calculators.[27] It would take months of testing small models on desk calculators before the air force team was convinced that the simplex method was both computationally efficient and practical, at least within the promise of digital computing capacity, and that it was not worthwhile to pursue better algorithms.[28] The IBM punch-card electrical accounting calculators available in 1948 and 1949, however, were not up to the task of manipulating the large rectangular matrices required for computing the optimal programming of Operation Vittles, much less for the larger wartime and peacetime programs for overall air force operations that would require the solution of simultaneous equations systems consisting of over 1,000 equations in 1,000 or more unknowns. In their 1949 report to the air staff titled *Machine Computation*, Wood and Geisler explained their recourse to what Herbert Simon would later label as "procedural rationality":

> Because the computational requirements are limiting in the development of alternative and optimum programs, we are engaged in research on mathematical procedures for simplifying and speeding up the computations of programs to facilitate effective utilization of new electronic computing equipment as it becomes available.[29]

The primary mathematical procedure for simplification to ensure computation that Project SCOOP resorted to was to use a "triangular procedure" to structure the relationships between end items and production activities. As with a fractal pattern, at almost every scale of modeling, Project SCOOP was forced to forgo an optimizing linear programming protocol in favor of the triangular procedure that was in essence the "rationalization, systemization, and mechanization" of the staff procedures highlighted in figure 2.5.[30]

> In order to obtain consistent programming the steps in the schedule were so arranged that the flow of information from echelon to echelon was only in one direction: thus the time phasing of information availability was such that the portion of the program prepared at each step did not depend on any following step. In our machine procedure we have similarly ordered the work in a series of stages.[31]

On the grandest scale of preparing for World War III, the SCOOP triangular patterning started with a war plan based on strategic guidance from the top echelon of the Department of Defense. The next step was to frame the air force as comprising distinct production activities and construct an input/output table of items and USAF activities during a war that would accomplish the strategic goals of the war plan. This wartime program would yield a quantitative statement of items (trained crews, equipment, bases) required to be on hand for M-day (mobilization day): "having defined this required M-day position, we may then determine the action necessary to proceed from our present status to the required M-day position under peacetime budgetary, personnel, and other limitations. This is the peacetime operating program."[32]

The construction of the input/output table of the civilian economy for the peacetime program required help from several federal government offices, including the Bureau of Labor Statistics and the Bureau of the Budget, as well as universities, including Harvard and the Carnegie Institute of Technology. The limitations on computing capacity were such that within each of the large-scale wartime and peacetime programs the unidirectional echelon-to-echelon approach would have to be repeated until one got to a level where mechanical computation with existing calculating resources was possible. The hope was that as computers improved, less and less triangular framing would be necessary and opportunities for maximization would become less localized. Before the UNIVAC arrived, however, Project SCOOP had to rely heavily on the triangular model even for relatively focused operations such as the Berlin Airlift.

2.5. Programming for Operation Vittles

In December 1948 Marshall Wood and George Dantzig presented a simplified version of their linear program for Operation Vittles at the winter meeting of the Econometric Society in Cleveland.[33] This model consisted of an input-output matrix and a dynamic equation system. The cells of the matrix contained coefficients that quantified the row items (equipment, personnel, and supplies) going into and coming out of the column activities (such as flying aircraft, training crews, and resting weary crews) over a three-month period of the airlift. For the longer planning periods that the model was intended for, all of the activity columns would be repeated to the right of the table making it quite "rectangular." The equation system included an objective function stating that the total costs summed over four three-month time periods and over all activities must be at its minimum value. The SCOOP team also demonstrated the dual nature of linear programming by presenting a version of the model with the total tonnage delivered to Berlin at its maximum. Successful computation would yield a "program" of action for the military in the form of quantities of the different types of activities needed to be performed at scheduled times in order to achieve maximum tonnage over twelve months (in the simplified version) or thirty-six months of the airlift, subject to resource constraints on the availability of the inputs and technological constraints in production.

Wood and Geisler presented a more formal elaboration at the seminal June 1949 conference on activity analysis organized by the Cowles Commission under the Commission's contract with the RAND Corporation and the USAF for research into the theory of resource allocation. Koopmans organized the conference to bring together operation researchers from the military as well as academic economists and mathematicians. The proceedings published in 1951 served for many years as a canonical text on both linear programming (what Koopmans called "activity analysis") and game theory.[34] Members of the Project SCOOP team presented seven of the papers, with Dantzig listed as an author or coauthor for five of them.

In their 1949 presentation, the SCOOP team claimed that one of the key advantages of the new mathematical model and algorithm they were presenting was its capacity to take into account dynamic opportunity costs: a cost of delivering more food today to Berlin is the opportunity foregone for ensuring greater food deliveries three months from now with a delivery today of material for constructing a runway. The preoptimization way of programming military operations had been plagued with an inability to consider alternative courses of action for determining the *best* program:

So much time and effort is now devoted to working out the operational program that no attention can be given to the question whether there may not be some better program that is equally compatible with the given conditions. It is perhaps too much to suppose that this difference between programs is as much as the difference between victory and defeat, but it is certainly a significant difference with respect to the tax dollar and the division of the total national product between military and civilian uses.

Consideration of the practical advantages to be gained by comparative programming, and particularly by the selection of "best" programs, leads to a requirement for a technique for handling all program elements simultaneously and for introducing the maximization process directly into the computation of programs. Such a technique is now in prospect.[35]

There are three key points that Wood and Geisler addressed in this passage: the promise of optimization to account for alternative uses and achieve the best outcome,[36] the declaration that this would enable the military to effectively pursue new operations without demands for higher taxes or lower civilian production, and the admission that the heralded innovation was "in prospect." With regard to the latter, Wood and Geisler had to present two hypothetical sets of input/output coefficients and respective derived equations to illustrate models for programming the Berlin Airlift: the optimizing "rectangular" model in prospect and the nonoptimizing "triangular" model (see figure 2.6) that could be solved on their current punch-card electrical accounting equipment.

The air force lacked the computing capacity to deal with the large rectangular matrices in Project SCOOP's original optimizing Berlin Airlift model. For several years to come, they had to rely on nonoptimizing models that were essentially small rectangular sets of a few equations arranged in such a way that their coefficient matrices formed a sequentially descending diagonal that bisected the matrix into triangles. The triangular model, or TriMod, as the SCOOP team often called it, rearranged and decomposed the general problems into hierarchical steps such that the algorithm had to solve for the levels for each activity in the earlier time period in order to solve for the activities at a particular moment.

Wood estimated that with the triangular model and punch-card calculators a full-scale wartime program would still take about six months to fully program, but the triangular protocols did yield computable solutions with significant cost reductions compared with previous planning practices. The triangular model with mechanical computation ensured consistent planning, smoother production, reduced waste and storage costs,

Triangularization of Berlin Airlift Model

Item	In/Out	Airlift Flying $x_1^{(t)}$	Resting Weary Crews $x_2^{(t)}$	Training New Crews $x_3^{(t)}$	Procuring Aircraft $x_4^{(t)}$
Supply Shipped by Airlift	In	−1			
	Out				
Weary Crews	In		1		
	Out	125			
Active Crews	In	130		.05	
	Out		1	1.00	
Aircraft	In	50		0.06	
	Out	49		0.05	1
Money	In	9,000	5	10.00	200

Equations

$$(1) \quad \alpha_{1,0}^{(t)} = x_1^{(t)},$$
$$(2) \quad 125x_1^{(t-1)} = x_2^{(t)},$$
$$(3) \quad \alpha_{3,0}^{(t)} + 130x_1^{(t)} - x_2^{(t-1)} + 0.55x_3^{(t)} = x_3^{(t)},$$
$$(4) \quad \alpha_{4,0}^{(t)} + 50x_1^{(t)} - 49x_1^{(t-1)} + 0.06x_3^{(t)} - 0.55x_3^{(t-1)} = x_4^{(t)},$$
$$(5) \quad 9,000x_1^{(t)} + 5x_2^{(t)} + 10x_3^{(t)} + 200x_4^{(t)}x = \text{money required in } i\text{th period};$$

where $\alpha_{1,0}^{(t)}$ = program of tonnages to be delivered in t = 1, 2, 3, and 4 (i.e., 1.5, 1.6, 1.8, and 2.0); $\alpha_{3,0}^{(t)}$ = inventory of crews initially available for airlift (i.e., 200, 0, 0, 0); $\alpha_{4,0}^{(t)}$ = inventory of aircraft initially available for airlift (i.e., 25, 0, 0, 0).

Figure 2.6. Project SCOOP's hypothetical triangular model with input-output coefficients and the equation system for determining resource requirements for quantitative goals in the Berlin Airlift. (Marshall K. Wood and Murray A. Geisler, "Development of Dynamic Models for Program Planning," *Activity Analysis of Production and Allocation: Proceedings of a Conference*, edited by Tjalling C. Koopmans [New York: John Wiley & Sons, 1951], 203.)

and increased analytical capacity for assessing feasibility. Indeed, the new accounting discipline and the rationalization of planning and data collection procedures that the triangular model engendered fulfilled many of the comptroller's aspirations, and air force reliance on variations of the triangular model persisted long after Project SCOOP ended.

The temporally constrained hierarchy of the triangular model could not incorporate dynamic considerations of opportunity costs of alternative production activities and it could not guarantee an optimum solution to qualitatively stated objectives such as "maximize tonnage delivered subject to constraints." The triangular model, however, could compute the required supporting activities to achieve a specified tonnage.[37] In other words, the triangular model, with far fewer computational resources than the optimizing linear programming model would have consumed, could satisfy an assigned aspiration level with precise calculations of necessary quantities of items and activities:

> With this formulation we have been able to solve programming problems involving 100 activities and 36 time periods in one day by using present punched card equipment, obtaining answers which are realistic and useful. In the more general formulation this would be represented by 3,600 equations in 3,000 unknowns.[38]

Ultimately, the Berlin Airlift did more for Project SCOOP than the SCOOP team did for Operation Vittles. Even with the triangular model the SCOOP team was not able to contribute to day-to-day planning of the airlift. The airlift, however, provided empirical and conceptual feedback that enabled the research team to hone their model, and begin to construct templates for data reporting and a working database of input/output coefficients for air force operations. The experience tested and proved the principles and concepts of planning by machine computation and provided a good instructional example to demonstrate the linear programming model to the air staff as well as academic economists.[39] It also opened a door, which would prove difficult to close, for making do with nonoptimizing or suboptimizing triangular models.

It was the promised economic rationality of optimization, however, that was celebrated at the 1949 conference where Project SCOOP presented its linear technology and the Operation Vittles model to the academic world. Tjalling Koopmans, the conference chair, even went as far as to claim that Dantzig's linear program model and simplex algorithm settled an earlier debate as to whether centrally planned allocation could lead to rational outcomes. European economists had engaged off and on since the late nineteenth century in the "socialist calculation" or "economic calculation" debate. The debate heated up with the perceived success of some forms of state production planning during World War I. In the 1920s and 1930s free-market champions such as Ludwig von Mises and Friedrich Hayek, however, had argued that the challenge of economic calculation prohibited planned economies from achieving an efficient allocation of resources and in so doing precluded rationality:

> Without economic calculation there can be no economy. Hence, in a socialist state wherein the pursuit of economic calculation is impossible, there can be—in our sense of the term—no economy whatsoever. In trivial and

secondary matters rational conduct might still be possible, but in general it would be impossible to speak of rational production any more. There would be no means of determining what was rational, and hence it is obvious that production could never be directed by economic considerations. What this means is clear enough apart from its effects on the supply of commodities. Rational conduct would be divorced from the very ground which is its proper domain. Would there, in fact, be any such thing as rationality and logic in thought itself? Historically, human rationality is a development of economic life. Could it then obtain when divorced therefrom?[40]

Koopmans consciously took up this challenge and argued that economic calculation and rationality in centralized allocation was now possible: "To von Mises' arguments regarding the unmanageability of the computation problems of centralized allocation, the authors oppose the new possibilities opened by modern electronic computing equipment. . . . Dantzig's model is an abstract allocation model that does not depend on the concept of a market."[41] What had been made clear to participants at the conference was the mathematical duality of maximizing output and minimizing costs and that the process of solving for the maximum output subject to constraints yielded what operation researchers called "efficiency prices" ("imputed prices," "accounting prices," or "shadow prices") that signaled worth in the absence of markets. This held out the prospect of computing meaningful valuations for planning the optimal allocation of resources in a system.[42]

The Cowles Commission report *Rational Decision-Making and Economic Behavior*, which announced the publication of the proceedings of the 1949 conference and detailed subsequent Cowles research studies, illustrated the commission's new emphasis on economic calculation and rationality:

> It was J. R. Hicks, the Oxford economist, who said that the foundations of economic theory were, essentially, nothing but "the logic of choice." Charles Hitch, of The RAND Corporation, expressed this in another way: "Economics is about how to economize." To be economical is to choose the best use of limited opportunities and resources. . . . All these cases of "economical" decision-making have the same logical content. In mathematical language, their common problem is to "maximize, subject to given conditions." "Rational behavior" and "optimal behavior" are still other words for economical decision-making.[43]

The report, issued two years after the conference on activity analysis, also entertained the notion that normative studies prescribing rational behavior had to take into consideration that actual behavior could be imperfectly rational or even irrational.[44] With their statement, "In order to make rational recommendations on human institutions and policies it is necessary to predict as well as we can people's actual, possibly irrational behavior," the Cowles Commission acknowledged a formal interest in research studies on less than rational behavior. Herbert Simon, who at the time of his participation in the activity analysis conference was working simultaneously under an air force research contract with Project SCOOP and the Cowles/RAND contract on resource allocation, would soon answer this call.[45] Although Simon would end up theorizing on imperfectly rational actual behavior, his starting point would be a normative perspective shaped by the frustrated computation of optimal solutions for US military planning. The mathematical programming of Operation Vittles would not be the only endeavor where limited computational capacity forced operations researchers to make do rather than maximize, and it was from his own confrontation with a noncomputability dragon that Simon would construct his theories of bounded rationality.

2.6. Project SCOOP and the Carnegie Institute of Technology

As part of their development of a large, peacetime interindustry mathematical model for determining the feasibility of M-day plans, Project SCOOP and the Bureau of the Budget's Division of Statistical Standards in 1949 awarded the Graduate School for Industrial Administration (GSIA) at the Carnegie Institute of Technology in Pittsburgh, Pennsylvania, a three-year grant for research on "intra-firm planning and control."[46] The commitment involved researching the upstream data generating process that fed into Project SCOOP's input/output models, improving production planning through an applied mathematics that combined logical, accounting, engineering, and computational frameworks, and training staff in new analytical methods of planning and optimization. This foundational contract between the military, the executive branch, and the university was a key path by which linear programming spread to operations research in private industry and by which management science became a profession backed by an analytical graduate business school curriculum and a professional organization.

To complement Project SCOOP's development of a peacetime interindustry model, the Carnegie group agreed to direct their intrafirm analysis of production planning to companies that were either very important to the economy or representative of a key industry. The Carnegie team had an additional mandate to ultimately make the analysis "operational" or "handbook ready." Their goal was to get down to a protocol so accessible that production managers with little mathematical training could routinely manipulate the decision rules by plugging in values for the unknowns in a simple equation. As was common in military-funded Cold War operations research projects, the mathematical analysis was meant to culminate in computable decision rules for the optimal allocation of resources. As was also common in such projects from the late 1940s to the 1970s, limited computational resources necessitated a rationalization of procedures and stimulated the development of new modeling strategies.

The faculty working on the GSIA air force research project split into two teams with "polar philosophies of optimizing behavior": the planning behavior approach of linear programming and the adaptive behavior approach of "servo control."[47] Each team ended up developing new procedures and strategies to match the optimization goal with the demands for accessible mathematics and effective computation.

GSIA economist William Cooper, CIT mathematician Abraham Charnes, and an engineer at the Philadelphia Gulf Oil Refinery, Bob Mellon, used linear programming to derive the optimum blend of aviation fuel at the refinery and presented their results at the 1951 Project SCOOP symposium on linear programming.[48] In order to make the linear programming approach "handbook ready" for production managers, Charnes developed a general means of handling complex mathematical problems that, in combination with Dantzig's simplex algorithms, made it possible to completely routinize the computing process.[49] His innovation for handling degeneracy opened the door for widespread industrial applications of linear programming, as did Charnes's and Cooper's publication of their GSIA lecture notes that served as the go-to text on linear programming for operations researchers for many years.[50] Armed with operational optimization tools, the GSIA team was at the forefront of professionalizing management science. Cooper was the founding president of The Institute of Management Science (TIMS), the first TIMS national meeting was held in Pittsburgh in October 1954, and in the first few years of the organization Charnes and Simon served as national officers on the editorial board of the professional organization's journal, *Management Science*.

At the other philosophical pole of optimization, Simon and engineer-turned-economist Charles Holt appropriated mathematical approaches to modeling the feedback loops in servomechanisms.[51] They hoped to leverage this emphasis on modeling adaptive behavior to determine a firm's optimal response to changing external information from the market. A specific aim was to derive decision rules for scheduling production for the White Motor Company truck assembly plant that would minimize manufacturing and inventory costs. The servo-control approach they used relied on mathematically accessible linear differential equations with constant coefficients. Simon asked, "Under what conditions can the optimal production paths be represented by such equations?"[52] Simon's answer was that the cost function to be minimized must have a quadratic form—it must consist of only linear or squared terms, but no higher-order (e.g., cubed) terms. With the transfer of their planning research projects initiated under Project SCOOP to Office of Naval Research (ONR) projects, Simon and Holt continued to explore the significance of the procedural economy gained by making the minimization of a quadratic production and inventory costs function the optimization objective.[53]

Project SCOOP's contract with the GSIA ended in June 1953, but the research for the air force had laid the foundation for the larger and longer contract with the ONR for planning and control of industrial operations that began in the fall of 1952.[54] Under the ONR contract Simon, Holt, the economist Franco Modigliani, and the engineer-turned-industrial administration doctoral candidate John Muth constructed, tested, and applied *dynamic programming* to derive linear decision rules to achieve optimal production rates at the Springdale paint manufacturing plant of the Pittsburgh Plate Glass Company.[55] Similar to the situation Project SCOOP had confronted, the company and the GSIA team had to make do with punch-card calculators to compute the optimal solutions from models and algorithms designed for digital computers. The GSIA would be the first group at the Carnegie Institute to get a digital computer, but that would not happen until 1956. A key information-processing hurdle for the ONR Paint Project team arose from the need to take into account uncertainty into their multistage decision-making model. In order to plan production they needed to estimate future demand for different types of paint, but data collection and computation hurdles made it difficult to incorporate a probability distribution of future sales into the protocol without making some heroic assumptions about the independence of demand in one time period compared with another.

Herbert Simon solved this problem of noncomputability with his "certainty equivalence" theorem. Simon demonstrated that if the cost function to be minimized in a dynamic programming problem could be approximated by a quadratic function, then a single expected value of future sales (e.g., an average of past sales), rather than the entire probability distribution for forecasts, would be sufficient to quantify linear decision rules.[56] As

Holt explained in a 1953 ONR report, approximating the total costs with a quadratic function "yields linear decision rules which are mathematically simple to derive and computationally easy to apply."[57]

Simon took the lesson of making do with existing computational resources when crafting normative modeling strategies and applied it to the realm of positive, descriptive economics. He argued in many forums that economists should be incorporating these approximating heuristics into their models of how economic man makes decisions.[58] In a working paper first circulated at the RAND Corporation in 1953 and published in the *Quarterly Journal of Economics* in 1955, Simon asserted that "the task is to replace the global rationality of economic man with a kind of rational behavior which is compatible with the access to information and the computational capacities."[59] In that essay, Simon explored ways of modeling the *process* of rational choice that took into consideration computational limitations.

In his 1957 book *Models of Man*, Simon introduced the term "bounded rationality." He argued that consumers and entrepreneurs were "intendedly rational," but they had to construct simplified models of real situations that were amenable to effective computation. For Simon a key to simplifying the choice process and reducing computational demands was "the replacement of the goal of maximizing with the goal of satisficing, of finding a course of action that was 'good enough.'"[60] Simon explored theories of limits on the perfect rationality assumed by economists. These limits included uncertainty about outcomes of decisions, incomplete information about alternatives, and complexity that defied computation. Such constraints on the information-processing capacity forced economic actors and operations researchers to focus on the decision process by, for example, taking into account the costs of searching and computing information or replacing optimizing approaches with heuristic approaches that complemented a satisficing approach.[61] Simon was not assuming that decision makers were irrational; rather, he argued that the limits on their capacity for collecting and processing the information needed to make the best decisions to meet their goals forced a new focus on the process of reasoning and problem solving.

At a talk at Groningen University in September 1973 and in revisions circulated in 1974 and published in 1976, Simon clarified his emphasis on the problem-solving process by making the distinction between what he called "substantive" and "procedural" rationality. Substantive rationality was the achievement of the best outcome given an optimizing goal; the rational consumer achieving maximum utility or the rational producer achieving maximum profits. In contrast to the economist's emphasis on the choice *outcome* that a rational economic man made, the psychologist focused on the *process* of how decisions are made. Procedural rationality dealt with reasonable deliberation.[62] Simon illustrated the difference between the two with examples from linear programming. The solution to Stigler's diet problem was substantively rational. Thanks to the linear programming model, the simplex solution algorithm, and contemporary computing equipment, an optimal least cost solution meeting the nutritional goals had been achieved. The "traveling salesman problem" of finding the city-to-city route that would minimize traveling costs was one of Simon's examples of procedural rationality. Computable optimal solutions were only possible for trivial setups of the problem. For the more complex traveling salesman problems, operations researchers searched for computationally efficient algorithms that would achieve good, but not necessarily optimal, solutions.[63] Bounded rationality in the form of limited information processing capacity led to procedural rationality. As Simon spelled out in other essays, the search process for the best procedures to make do might itself involve optimization, for example, determining an optimal stopping point for the search process or minimizing the costs of the search process, even when the original optimization was precluded by limited computational resources.

Simon perceived "simplification of the model to make computation of an 'optimum' feasible" (as in approximating costs with a quadratic equation) and "searching for satisfactory, rather than optimal choices" (as in programming with a triangular input/output matrix) as examples of satisficing, rather than optimizing, behavior. The aim of both approaches was to construct "practicable computation procedures for making reasonable choices."[64] Simon acknowledged that a search for the most efficient method of approximating sometimes made it difficult to draw a formal distinction between optimizing and satisficing. He argued, however, that there was often a major practical difference in emphasis, and that the aspirational approach of searching for satisfactory choices could lead to better results than the first-approximate-then-optimize approach.[65]

The thrust of Simon's argument in many of his essays contrasting substantive and procedural rationality addressed positive economics that purported to describe actual behavior. Simon perceived consumers and producers as organisms with limited computational capacity. Therefore economists, Simon asserted, should learn from psychologists, as well as from their own discipline's experience with normative operations research, and focus more on the process of how decisions are made. As is evident in a symposium on economics and operations research in the May 1958 issue of the *Review of Economics and Statistics*, Simon was not alone in drawing this conclusion.[66] Several of the authors, including Simon's colleague William Cooper as well as Charles Hitch, Thomas Schelling, and Daniel Ellsberg from the RAND Corporation, spoke not only to how economists could contribute to improvements in operations research, but also to how their own operations research experience with approximation and good, alternative, nonoptimizing rules should be incorporated into microeconomic theory. It was Simon, however, who provided clarity with the naming of "bounded" and "procedural" rationality.

2.7. Programming after Project SCOOP

Significant Department of Defense budget cuts accompanied the new policies of Dwight Eisenhower, inaugurated as president of the United States in January 1953, and the end of armed Korean conflict in July of that year. With a staff of fifty and major research operating expenses, Project SCOOP was vulnerable to such cuts. Also *planning*, particularly planning the entire US economy as was Project SCOOP's ambition in its interindustry modeling for their peacetime program, was taboo under the new secretary of defense, Charles E. Wilson. In the fall of 1953 the air force disbanded Project SCOOP, formally acknowledging an end to the early development stage of mathematical programming and a commitment to a new stage of implementation, albeit now confined to modeling only air force activities. The air staff also changed the name of the Planning Research Division to the Computation Division.[67]

Mathematical programming, however, continued to thrive both at the comptroller's office at the Pentagon and at the RAND Corporation, to where George Dantzig and Murray Geisler had migrated. Even after the arrival of the UNIVAC, the air force had to rely heavily on the triangular model for programming operations with thousands of activities. Former SCOOP members, including Walter Jacobs and Saul Gass, worked at the Pentagon to design models and code algorithms that could be solved with existing computers and be operationally friendly to those having to formulate myriad air force routines. In 1956 they replaced the triangular model with the Trim model (also square and non- or suboptimizing) that they had designed "as a production system" that disciplined and mechanized the data reporting process from various air force departments.[68] The Trim model was used to construct budgets and to determine for any specific war plan the monthly requirements for bombs, ammunition, fuel, personnel, and so forth. In cases of smaller military operations, linear programming on the UNIVAC with rectangular optimizing models was possible by the mid-1950s.[69]

So far in this chapter we have neglected the other superpower engaged in the Cold War. If necessity is the mother of invention, why didn't Soviet requirements for planning the entire national economy spur an early development of linear programming there? There was a major, albeit neglected, development in the linear technology for optimum resource allocation in 1939. In his capacity as a consultant to a plywood enterprise, Leonid Kantorovich, professor of mathematics at Leningrad State University, was confronted with the economic problem of allocating raw materials in order to maximize equipment production subject to constraints. He formulated a linear programming model and suggested an iterative solution process similar to, but not identical to, the simplex method that held out the prospect for calculating "resolving multipliers" (the "efficiency prices" or "shadow prices" of Dantzig's model). In May 1939 Kantorovich made two presentations of his new mathematical approach to the Institute of Mathematics and Mechanics of the Leningrad State University and to the Leningrad Institute for Engineers of Industrial Construction. That same year the Leningrad University Press printed his booklet, *The Mathematical Method of Production Planning and Organization*. A lack of computational capacity in the early 1940s, a distrust of a mathematical approach to planning allocation of resources, and the preoccupation of war with Germany led to the neglect of Kantorovich's contribution to scientific management in the USSR. In the late 1950s, planners and academics in the USSR began to acknowledge the usefulness of Kantorovich's protocol and its excellent fit with the growing Soviet interest in cybernetics.[70] In 1975, the Swedish Nobel Committee awarded Kantorovich and Koopmans the Sveriges Riksbank Prize in Economic Sciences for their contribution to the theory of the optimum allocation of resources. To the consternation and anger of many, including Koopmans, Dantzig was not included in the honor.

In June 1952, Dantzig left Project SCOOP to continue his developmental work for the air force at the RAND Corporation. There Dantzig worked with William Orchard-Hays to improve the computational efficiency of the simplex algorithm, to adapt it to new computers, and to develop commercial-grade software for solving linear programs. Looking back on that work with Dantzig, Orchard-Hays described their occupation: "An algorithm designer is an engineer who works with abstract concepts rather than physical materials. The designer's goals are efficiency and that the algorithm works; it should give correct results reliably for a class of problems."[71] The algorithm designers at RAND were constructing what Simon named *procedural rationality*, "the rationality of a person for whom computation is the scarce resource."[72]

2.8. The Rise of Procedural Rationality

In the 1958 symposium on Economics and Operations Research the economist Daniel Ellsberg described the conditions in the Cold War military sphere that made "problems of choice and allocation almost unbearably pressing."[73]

> The budget famine, the sudden military challenge, the unprecedented variety of alternative weapon systems with their long lead-times, rapid obsolescence, high cost, and excruciatingly technical claims: these basic pressures on the Secretary of Defense are dramatized by the inter-service rivalry with the public for funds and resources and with each other for allocations and control of weapons.[74]

That description applied equally well to the earlier decade in which the blockade of and airlift for the western sectors of Berlin increased the urgency of US Air Force attempts to compute optimum programs of action. Reminiscing on this task forty years later, George Dantzig described the conditions for the development and rapid diffusion of a mathematical protocol that held out the novel prospect of prescribing a rational allocation of resources in a complex system: "The advent or rather *the promise* that the electronic computer would exist soon, the exposure of theoretical mathematicians and economists to real problems during the war, the interest in mechanizing the planning process, and last but not least the availability of money for such applied research all converged during the period 1947–1949. The time was ripe."[75] Dantzig claimed that "*True optimization* is the revolutionary contribution of modern research to decision processes." With the linear programming model and the simplex solution algorithm, Dantzig and his air force colleagues had initiated "a vast *tooling-up*" that would characterize management science for decades. He acknowledged, however, that "computation of truly *optimum* programs was of course beyond the original capabilities of SCOOP."[76] For Operation Vittles and most other air force operations planned in its five-year existence, Project SCOOP made do with satisficing rather than optimizing protocols.

By the early 1950s operations researchers had the linear programming tool for modeling the optimal allocation of resources across activities and the dynamic programming tool for modeling the optimal allocation of resources across time periods. The essential link of mathematical programming to coding for digital computers was accompanied by the standardization of model formulation and algorithmic design, inductive and deductive time studies of computational procedures, and deskilling goals applied to the data input process as well as the rule-based decision output process. Cold War military specifications for operational numerical solutions to calculation-for-allocation problems forced researchers to confront the bounds of their intended rationality and to direct reasonable deliberation to procedures for achieving good-enough, if not the best, decision rules.

Simon wanted to take this lesson from the normative, prescriptive realm of management science and apply it to descriptive realm of positive economics. According to Simon, economists had made valuable contributions to social science with their explicit and formal assumption of rationality, but they had neglected to take into account the limitations on the computing capacity of the consumers and entrepreneurs that they modeled. For Simon, the economists' representation of rationality as an optimal *outcome* for a goal-oriented decision maker was unrealistic and incomplete. Simon's emphasis on procedural rationality begged for a psychological focus on the *process* of problem solving. As we will see in the next chapter, psychologists channeled their analytical emphasis on the decision-making process to generating an understanding of the conditions for the de-escalation of tensions surrounding another Cold War blockade—Cuba 1962.

THREE

Saving the Planet from Nuclear Weapons and the Human Mind

After almost two weeks of cloudy weather, a U-2 reconnaissance flight over Cuba, located just ninety miles off American shores, finally managed to obtain some photographs. On Sunday, October 14, 1962, it brought them back. A day later, the National Photographic Interpretation Center examined the developed film and found the early stages of construction of a medium-range ballistic missile base (figure 3.1). A missile launched from such a site could easily threaten millions in any major city on the East Coast of the United States, including the capital, Washington DC. From Tuesday, October 16, the "thirteen days" of the Cuban Missile Crisis unfolded, bringing the world the closest it has ever been to a thermonuclear exchange between the two superpowers.[1] The crisis as it has been interpreted in the historical literature is particularly "American," as this was the side that was newly targeted by offshore missiles (the Soviet Union had been ringed with nuclear missile bases, most significantly in Turkey, for some time). The deliberations within the administration of John F. Kennedy about how to resolve the situation without nuclear escalation have since become one of the most studied episodes in twentieth-century history.

Kennedy assembled an ad hoc group of advisors, the Executive Committee (universally known as ExComm), composed of foreign policy and defense specialists (like former secretary of state Dean Acheson and his own secretary of defense Robert S. McNamara), personal advisors (like Ted Sorenson), and his brother, attorney general Robert F. Kennedy. In order not to arouse public concern (or Soviet preemptive action), Kennedy kept to his official schedule, including campaigning for congressmen running in the upcoming midterm elections. Meanwhile, the ExComm met for long stretches, debating possible responses to the Cuban situation. One thing was clear: those missiles had to go. How to accomplish this was another

Figure 3.1. One of the first U-2 images showing construction
of a missile base in Cuba, dated October 14, 1962.

matter: through a preemptive air strike to destroy the bases, through a blockade, through a trade of missiles? On Sunday, October 20, Kennedy returned from a campaign trip to the Midwest and resolved with the Ex-Comm to opt for a naval quarantine of Cuba. The United States Navy stopped its first Soviet-chartered freighter on October 26, searched for military contraband, and passed it through as clean. After tense exchanges of letters between Soviet premier Nikita S. Khrushchev and Kennedy, the former agreed on October 28 that he would remove the missiles from Cuba, quietly acceding to a secret deal that Kennedy would decommission obsolete Jupiter missiles from Turkey. The quarantine officially ended on November 21. The world had stepped back from the brink.

Ever since, scholars have been trying to explain why. Given that this incident took place during the apex of the debates over Cold War rationality, it is not surprising that a dominant interpretation was to try to explain all the moves in the game as "rational." To do so, of course, one needs a definition of what rationality might mean. In one contemporary articulation dating from 1961, political scientist Sidney Verba characterized "rational" decision making in foreign policy according to a formula well-known from the calculating economic theory of choice we encountered in the previous chapter:

> Rational models of individual decision-making are those in which the individual responding to an international event bases his response upon a cool and clearheaded means-ends calculation. He uses the best information available and chooses from the universe of possible responses that alternative most likely to maximize his goals. . . . [T]he decision will either have no psychological side-effects on the decision-maker (he will not experience tension release or guilt because of it), or, if there are psychological side-effects, they will be irrelevant as far as the nature of the decision is concerned.[2]

Such frameworks were immediately deployed to explain what had happened in Cuba: to attempt to understand why orthodox deterrence theory had *failed* to prevent the Soviets from placing the missiles, and then how a moderate quarantine had *succeeded* in breaking the stalemate without escalation. Academic blood feuds have raged ever since. As political scientist Richard Ned Lebow has put it, "The Cuban missile crisis might be likened to a Rorschach test: the ink blots that constitute the few facts reveal little that is conclusive about Soviet policymaking, but the responses of political scientists to them say a lot about their anxiety concerning nuclear war."[3] The debate in that closed room—complete with secret tapes recording the deliberations—became what we will see formulated in the next chapter as a "strange situation": a controlled device to explore the rationality of human behavior. It just happened that this case was not contrived; the nuclear threat was real.

Due in no small part to the nail-biting tension of October 1962, during the several decades of nuclear-armed standoff between the United States and the Soviet Union, nuclear weapons were repeatedly cast, even by such a foreign policy maven as George Kennan—the most influential "Russia Hand" in the US government at the dawn of the Cold War and the author of the strategy of "containment" (although later a prominent critic of the same as actually implemented)—to be fundamentally *irrational* devices: "To my mind, the nuclear bomb is the most useless weapon ever invented. It can be employed to no rational purpose."[4] This kind of language was sometimes used even by nuclear strategists who certainly did not think the weapons themselves were devoid of rational uses. They were the ones who advocated the most broadly endorsed nuclear strategy, mutual assured destruction; for many, the acronym symbolized the atomic age. However

insane its conclusions might seem to the uninitiated, the esoteric specialization of nuclear strategy was among the most committed of the human
sciences to adopt and deploy rational-choice frameworks, game theory,
and systems analysis to its stated problem: the waging (or prevention) of
nuclear war.[5] At every moment, the various doctrines were engaged with
the struggles over rationality in the Cold War.

This chapter uses two case studies to explore some contours of just what
"rationality" meant in the nuclear realm from the late 1950s to the late
1970s. Both cases depart from Verba's articulated definition of rationality in
that they highlight precisely those features of the human psyche he called
"nonrational" and others would term "irrational." In both cases, it was the
goal of the theorists to show how those specific features prevented flesh-
and-blood humans from making "rational" decisions in foreign-policy
situations. Both of them—the strategy of unilateral disarmament proposed
by University of Illinois psychologist Charles E. Osgood (1916–1991),
which he dubbed GRIT (graduated and reciprocated initiatives in tension
reduction), and the notion of "groupthink" developed by his almost exact
contemporary, Yale psychologist Irving Janis (1918–1990)—implied ways
of harnessing the unavoidable irrationality of the human mind to defeat
itself. For Osgood, true rationality required reincorporating those features
of the mind that orthodox defenders of the tenets of Cold War rationality in nuclear strategy had stripped out; only by being aware of the limitations on rationality could you guide the mind into producing rational
decisions. For Janis, the case was simpler: group dynamics could stymie
rational thought, and instead of domesticating these irrational features to
rescue policymakers, he proposed quarantining them, like Cuba, until they
behaved properly. In both instances, psychology was brought to bear on
the core technologies of the Cold War—nuclear weapons—in the hope of
yoking them to the mind, and in some cases resulted in reinforcing the formulations of Cold War rationality they had set out to criticize.

3.1. Herman Kahn and His Chicken

Critics like Osgood and Janis had many causes for concern that the appeal of algorithmic, mechanical calculations would inevitably lead to the
self-evident catastrophe of nuclear war. The Cuban Missile Crisis did not
create their worries, but it very effectively dramatized them. The United
States ended World War II with a single atomic bomb undetonated—it was
not even assembled yet—and for a few years nuclear production remained
in the single or double digits as President Harry S. Truman attempted to

navigate both domestic and international postwar crises. Heedless of the mismatch, nuclear strategy proceeded apace, with military plans like HALF-MOON and BROILER in 1947 calling for the deployment of dozens, if not hundreds, of yet-nonexistent nuclear weapons.

The detonation of the first Soviet nuclear bomb on August 29, 1949, and the speedy American detection of that event precipitated a massive expansion of atomic infrastructure. The advent of the hydrogen bomb in the early 1950s (which not only exponentially increased the destructive power of nuclear weapons but enabled them to be miniaturized and placed atop missiles), the outbreak of the Korean War in June 1950, American worries about the costs of maintaining a conventional army to match the Soviets and thus increasingly relying on a nuclear-heavy strategy of "massive retaliation," the falling price of uranium as domestic prospecting glutted the market—all these prompted the production of ever more weapons, which soon were incorporated into plans: the Basic War Plan of the US Strategic Air Command in March 1954 called for up to 735 bombers to be launched against the Soviet Union in a massive coordinated attack, deploying a sizeable chunk of the 2,063 warheads then at the Americans' disposal (up from 369 in 1951); the number of targets to be destroyed was already 3,560 by 1960, using up most of the 20,434 warheads in the US stockpile. There were soon so many warheads—the peak was 32,193 in 1966—that planners sought targets to match their capability, rather than adjusting capability to match targets[6] (figure 3.2). The death tolls envisaged were in the tens of millions for Americans, and hundreds of millions globally.

One of the most potent metaphors of the nuclear arms race was introduced not by a crew-cut game theorist at RAND, but by Bertrand Russell, the liberal lion of philosophy. For years, Russell had worried about nuclear weapons; in July 1955 he released a manifesto coauthored with physicist Albert Einstein (who had died earlier that year) calling for disarmament, and their eponymous statement sparked intense international activism, not least prompting the formation of the Pugwash Conferences. Russell's metaphor came four years later and had a career just as distinguished as his manifesto. Here is the original presentation in his own inimitable style:

[Brinksmanship] is a policy adapted from a sport which, I am told, is practised by the sons of very rich Americans. This sport is called "Chicken!" It is played by choosing a long straight road with a white line down the middle and starting two very fast cars towards each other from opposite ends. Each car is expected to keep the wheels of one side on the white line. As they approach

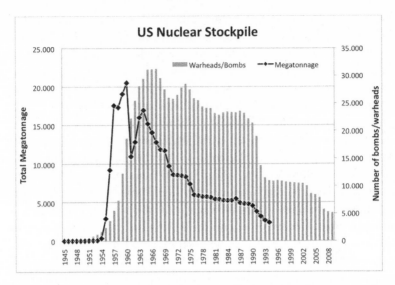

Figure 3.2. Graph comparing the historical evolution of both numbers of weapons and mega-tonnage in the US nuclear arsenal. The megatonnage declined as both superpowers shifted to smaller, independently targeted nuclear warheads designed for precise strikes on militar-ily valuable installations (a strategy known as "counterforce"). The large megaton weapons intended for city bombing were thus deemphasized, although not phased out entirely. (http://www.defense.gov/npr/docs/10-05-03_fact_sheet_us_nuclear_transparency_final_w_date.pdf and https://www.osti.gov/opennet/forms.jsp?formurl=document/press/pc26tab1.html.)

each other, mutual destruction becomes more and more imminent. If one of them swerves from the white line before the other, the other, as he passes, shouts "Chicken!," and the one who has swerved becomes an object of con-tempt. As played by youthful plutocrats, the game is considered decadent and immoral, though only the lives of the players are risked. But when the game is played by eminent statesmen, who risk not only their own lives but those of many hundreds of millions of human beings, it is thought on both sides that the statesmen on one side are displaying a high degree of wisdom and courage, and only the statesmen on the other side are reprehensible. This, of course, is absurd. Both are to blame for playing such an incredibly dangerous game. The game may be played without misfortune a few times, but sooner or later it will come to be felt that loss of face is more dreadful than nuclear an-nihilation. The moment will come when neither side can face the derisive cry of "Chicken!" from the other side. When that moment is come, the statesmen of both sides will plunge the world into destruction.[7]

And so the game of Chicken was added to the conceptual apparatus of nuclear thinking. We should stress one point here: the nuclear arms race was not straightforwardly Chicken, any more than it was intrinsically the prisoner's dilemma or any other game; rather, individuals had to mount intellectual arguments to make the arms race intelligible *as* Chicken (or as prisoner's dilemma), or to refute that characterization. And it was necessary for nuclear strategists to refute it, because—as Russell contended and many later commentators agreed—if the only sane way to behave in a game of Chicken was to refuse to play, then the game would imply a warrant for nuclear disarmament.

The most prominent respondent to Russell's challenge, and the man who formed the iconic representative of the Cold War rationality that psychologists like Osgood and Janis railed against, was Herman Kahn. Kahn is today perhaps the most widely known nuclear strategist from this period, although he was only one of many based at RAND, and of the second generation of nuclear intellectuals at that. (His influence was likely substantially less, in real policy terms, than Albert Wohlstetter's, for example.) There were many distinctive things about Kahn—his enormous girth, his Bronx accent, the ubiquity of his magnum opus *On Thermonuclear War* (1960), the role that ideas from the latter (especially the "Doomsday machine") played in Stanley Kubrick's *Dr. Strangelove*—but perhaps the most characteristic feature of Kahn within the sphere of nuclear theory was his commitment to rationality in conflict. Kahn vocally insisted that at each stage of the nuclear cycle, up or down the ladder of escalation, *all* the decisions could, should, and likely would be rational.[8] "Nor are the psychological processes of the players relevant" for Kahn, noted Anatol Rapoport (see section 5.2); Kahn believed that everything about nuclear war could be understood using the core principles of rational-choice theory.[9]

He expressed his views most crisply in the appropriately titled *On Escalation* (1965), which we will focus on here in lieu of *On Thermonuclear War*, which has been discussed extensively elsewhere.[10] Kahn presented escalation as a classic "ladder": one could go up or down the rungs, finely calibrating conflict to achieve the appropriate level of deterrence. At some moment in these fluctuations, one crossed the nuclear threshold—it was necessary to be willing to do this in order to render deterrence credible (as we will see with the case of Richard Nixon in chapter 5), and credibility was Kahn's central problematic. The ever-present potential for nuclear escalation was built into the system of deterrence as a precondition for its normal functioning. Kahn agreed with Bertrand Russell about the rational outcome—no nuclear war—but he disagreed about which rules to follow in reasoning one's way to that end. For Kahn, thinking through all the scenarios in the clear light of axiomatic, formalized rationality was the only way to prepare for the vicissitudes of conflict. Grandstanding and rhetoric (not to mention moralizing) had no part of it—although that did not stop Kahn from engaging in the first two with gusto, while denying the relevance of the third in matters of nuclear strategy.

That meant Chicken had to be addressed, for "we may have to be willing to play the international version of this game whether we like it or not." Kahn devoted several pages early in *On Escalation* in discussing this "useful—if misleading—analogy" that "greatly oversimplifies international conflict." Kahn conceded that the analogy provided some insight into the arms race, especially its symmetric character and the use of signaling (wearing dark glasses, throwing whiskey bottles out the window, visibly detaching the steering wheel to mark the inability to move the vehicle off course) to highlight one's determination. But, in the end, the comparison was not quite right:

> "Chicken" would be a better analogy to escalation if it were played with two cars starting an unknown distance apart, traveling toward each other at unknown speeds, and on roads with several forks so that the opposing sides are not certain that they are even on the same road. Both drivers should be giving and receiving threats and promises while they approach each other, and tearful mothers and stern fathers should be lining the sides of the roads urging, respectively, caution and manliness.[11]

The point for Kahn, and for Thomas Schelling, whose April 1963 lecture on this topic Kahn quoted at length, was that if Russell was right—that the only solution to Chicken is for both players to agree in advance not to play—then one needed to understand which situations might devolve into Chicken and preempt them. If we did not analyze the scenarios *rationally*, then we would fail to avert nuclear war. Schelling's example of a case when "nations are thrust into situations where both sides must show their mettle"?[12] The Cuban Missile Crisis.

3.2. Charles Osgood Shows GRIT

One of the most riveting documents produced during the Cuban Missile Crisis was the almost hysterical first letter from Nikita Khrushchev to John F. Kennedy on October 26, which pleaded with the American president to pull the world back from the nuclear brink. The next day, a calmer but more hard-line letter was received, prompting debate among the Ex-Comm over whether there had been a coup against the Soviet premier. The second letter demanded a trade of missiles from Turkey in addition to once more insisting on a guarantee that the Americans would not invade Cuba. Weighing the two letters against each other provided one of the central dramas of the ExComm. The first letter, which no one doubts was written personally by Khrushchev, merits a closer read. Amid the panic and paranoia, he inserted a vivid metaphor for the situation:

> If you have not lost your self-control and sensibly conceive what this might lead to, then, Mr. President, we and you ought not to pull on the ends of the rope in which you have tied the knot of war, because the more the two of us pull, the tighter the knot will be tied. And a moment may come when that knot will be tied so tight that even he who tied it will not have the strength to untie it, and then it will be necessary to cut that knot, and what that would mean is not for me to explain to you, because you yourself understand perfectly well of what terrible forces our countries dispose.[13]

This passage recognizes the capacity of nuclear crises to destroy self composure, and argues the arms race was mutually constituted from unilateral moves, so that only mutual cooperation could save the world. It almost reads like a précis of GRIT.

Although now it is almost entirely forgotten, Charles Osgood's GRIT was an important foray of the human sciences into nuclear strategy, the most prominent such intervention in the direction of *de-escalating* the arms race. GRIT was a core staple of the "conflict resolution movement" as it developed in the late 1950s, institutionalized in such places as the *Journal of Conflict Resolution* and its associated center at the University of Michigan.[14] It is a bit difficult to encapsulate GRIT in a nutshell, for—as one of its prime advocates would later claim—"GRIT is not a theory; it is a program, a technique."[15] Like many of the reformulations of rationality in the context of Cold War debate, it was more a method than a theory, an approach and not a synthesis. A general outline can be gleaned from this colorful image from Osgood's 1959 article that introduced the (not-yet-named) GRIT approach:

> Imagine two husky men standing facing each other near the middle, but on opposite sides, of a long and rigid, neatly balanced seesaw. As either man takes a step outward, the other must compensate with a nearly equal step outward on his side or the balance will be destroyed. The farther out they move, the greater the unbalancing effect of each unilateral step and the more agile and quick to react both men must become to maintain the precarious

equilibrium. To make the situation even worse, both of these husky men realize that this teetering board has some limit to its tensile strength—at some point it is certain to crack, dropping them both to destruction. So both men are frightened, but neither is willing to admit it for fear the other might take advantage of him. How are these two men to escape from this dangerous situation—a situation in which the fate of each is bound up with that of the other?

One reasonable solution immediately presents itself: let them agree to walk slowly and carefully back toward the center of the teetering board in unison. To do this they must trust each other. But these men *dis*trust each other, and each supposes the other to be irrational enough to destroy them both unless he (ego) preserves the balance. But now let us suppose that, during a quiet moment in the strife, it occurs to one of these men that perhaps the other really is just as frightened as he is and would also welcome some way of escaping from this intolerable situation. So this man decides to gamble on his new insight and calls out loudly, "I am taking a small step *toward* you!" The other man, rather than have the precarious balance upset, also takes a step forward, whereupon the first takes yet another, larger step.[16]

Charles Osgood was a psychologist, and a rather famous one at that. A graduate of Dartmouth College (class of 1939), he did his doctoral work in psychology at Yale under the doyen of learning theory, Clark L. Hull; became president of the American Psychological Association in 1963; and was elected to the National Academy of Sciences in 1972. In arms control, he supplied position papers to Eugene McCarthy and George McGovern and served as a consultant to the air force, the navy, and eventually the Arms Control and Disarmament Agency, which was founded in 1961 by President Kennedy at the suggestion of Senator Hubert H. Humphrey in response to the pressure of the ban-the-bomb movement. After some war work training B-29 gunners and a teaching stint in Connecticut, he moved to the University of Illinois at Urbana-Champaign in 1949, where he remained until 1982, when he was forced into retirement by the onset of Alzheimer's disease.[17]

His career was built around rigorous research in psycholinguistics, culminating in the development of the "semantic differential" technique, which remains one of the field's essential tools to this day. The purpose of the semantic differential was to resolve a question that had motivated Osgood since his undergraduate days: how does one measure meaning? In 1952, shortly after moving to Illinois, Osgood began to scale up his model for measuring national stereotypes from the war years.[18] In that earlier project, Osgood had created a series of polar qualities (e.g., "cruel/kind"), and set them at opposite ends of a seven-point scale. Respondents ranked each national group (e.g., "Japanese") on the scale. The net result showed tremendous consistency of social perceptions both within individual subjects and across a social group. The semantic differential is this method on steroids. As Osgood and his coauthors described it in their monumental unveiling of the theory, *The Measurement of Meaning*, "The semantic differential is essentially a combination of controlled association and scaling procedures. We provide the subject with a concept to be differentiated and a set of bipolar adjectival scales against which to do it, his only task being to indicate, for each item (pairing of a concept with a scale), the direction of his association and its intensity on a seven-step scale."[19] The concepts to be tested were more general ("mother," "tornadoes") and the qualities standardized (such as "hard/soft"). Once the results came in, Osgood and his collaborators analyzed them using multifactorial analysis on the massive (and conveniently-located) ILLIAC computer, generating a huge database that enabled comparison of meaning between or within cultures. Experimental psycholinguists elaborated and disseminated the tool widely.[20]

In the mid-1950s, while engrossed in this research program, Osgood stumbled into politics. The escalation of McCarthyism and Red-baiting kindled the embers of his liberalism, and he started reading the newspaper and getting more excited—and worried—about the politics of the Eisenhower years. Most of this engagement was passive until he went on sabbatical to the Center for Advanced Study in the Behavioral Sciences in Palo Alto, California, in 1958. In the next office he found a fellow psychologist, Jerome Frank, who had already become very involved in the antinuclear movement. Osgood began thinking about the arms race, and eventually published a version of the theory that would become GRIT in a paperback collection called *The Liberal Papers*.[21] From this point on, Osgood vigorously lobbied for arms control, with an intensity that only increased after President Lyndon Johnson escalated the war in Vietnam. Osgood abandoned everything but teaching and dedicated himself to spreading the word.[22]

Osgood published his first article on the GRIT idea in 1959, and he continued to develop it over the decades that followed, yet the fullest expression of the notion remains his 1962 book *An Alternative to War or Surrender*, where he noted that "the essential idea of GRIT" is "that the tension/arms race spiral may offer the model for its own reversal." He continued:

As a type of international behavior, just what is an arms race? *An arms race is a kind of graduated and reciprocated, unilaterally initiated, internation action. . . .*

But an arms race is obviously a *tension-increasing* system; it is a spiral of terror. By reversing one of the characteristics of an arms race, we may be able to transform it into a spiral of trust. This would be a graduated and reciprocated, unilaterally initiated, internation system that was *tension-decreasing* in nature.[23]

GRIT was intended to be the functional opposite of Herman Kahn's theory of escalation. If for Kahn the central problem was deterrence's credibility, for Osgood it lay much deeper: the inability of the human mind to function according to the seamless rationality Kahn posited. Osgood insisted that "deterrence is more a psychological question than a technological answer."[24] He started from a heuristic he dubbed "the Neanderthal Mentality," which boiled down to a claim that the core of our instinctive emotional reactions were rooted in the Stone Age, and thus "our facility with technology has completely outstripped our ability to understand and control ourselves."[25] These ingrained emotional reactions meant that while Kahn might think that "*physically*, escalators can run down as easily as up, but, *psychologically*, it is much easier to keep on going up than to stop and back down."[26]

The root difficulty was tension. For Osgood, weapons not only made people tense; the arms race made humans positively irrational:

Emotional tension produces stereotypy in our thinking; it restricts the range of alternatives we can conceive of, rendering us more rigid than flexible, and shortens our perspective, substituting blind reactions to immediate pressures for long-term persistence toward ultimate goals. In other words, paradoxically, the psychological conditions of prolonged deterrence produce the very states of mind which make it harder and harder to maintain deterrence.[27]

Or, as he put it in 1963, "as emotional stress levels increase beyond some optimum level, certain nonrational mechanisms in human thinking become more prevalent," making it less likely that "political man [will behave] rationally"[28] (see figure 3.3).

Thus Kahn's central assumption was false. Osgood concluded that an accurate psychological portrait was necessary: "Only to the extent that we understand the workings of our own minds as they grapple with a complex reality can we hope to achieve rationality in making decisions."[29] If the ideal type of Cold War rationality could be understood as a corrective to human irrationality, Osgood turned the tables: our caveman psychology

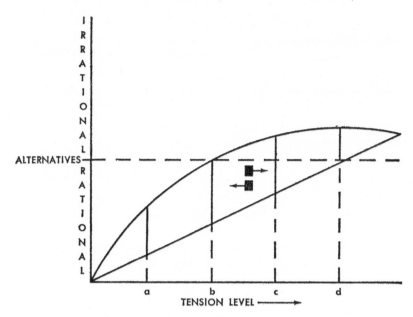

Figure 3.3. Osgood's schematic representation of the relationship between tension and rationality (Charles Osgood, *An Alternative to War or Surrender* [1962; Urbana: University of Illinois Press, 1970], 60.)

was inevitably part of us, and thus we needed to modify the algorithmic rigor of the rationality defended by Kahn and his ilk.

Hence GRIT: nation X decides that the arms race with nation Y is no longer tolerable. Nation X announces unilaterally that it will undertake a series of conciliatory moves (P, Q, R, and S, such as eliminating this weapons system, closing that foreign base, allowing its scientists to attend a previously boycotted conference) within the next year *regardless* of whether nation Y reciprocates (although it is explicitly invited to do so). Nation X implements P, Q, R, and S on schedule, even if nation Y responds mildly aggressively to these moves. By fulfilling its conciliatory promises, Osgood maintained, the cycle of escalating mistrust would be broken, and in most instances nation Y would respond with its own series of announced unilateral conciliatory moves. This cycle should repeat itself for some time, until enough armaments are eliminated and enough trust is established to allow for formal negotiations to succeed. "My colleagues in psychology," Osgood noted, "would recognize this strategy as the familiar process of deliberately 'shaping' behavior, here being suggested for use on an international scale."[30]

But GRIT was not reckless. Osgood was quite adamant that any situation in the nuclear age was risky—not least the status quo—and thus while GRIT involved some level of risk due to exploitation by the opponent (i.e., the Soviets), if one chose one's unilateral initiatives carefully, according to the following set of rules, one could make GRIT the *least* risky alternative:

RULES FOR MAINTAINING SECURITY

1. Unilateral initiatives must not reduce one's capacity to inflict unacceptable nuclear retaliation should one be attacked on that level.
2. Unilateral initiatives must not cripple one's capacity to meet conventional aggression with appropriately graded conventional response.
3. Unilateral initiatives must be graduated in risk according to the degree of reciprocation obtained from an opponent.
4. Unilateral initiatives should be diversified in nature, both as to the sphere of action and as to geographical locus of application.

RULES FOR INDUCING RECIPROCATION

5. Unilateral initiatives must be designed and communicated so as to emphasize a sincere intent to reduce tensions.
6. Unilateral initiatives should be publicly announced at some reasonable interval prior to their execution and identified as part of a deliberate policy of reducing tensions.
7. Unilateral initiatives should include in their announcement an explicit invitation to reciprocation in some form.

RULES FOR DEMONSTRATING THE GENUINENESS OF INITIATIVES AND RECIPROCATIONS

8. Unilateral initiatives that have been announced must be executed on schedule regardless of any prior commitments to reciprocate by the opponent.
9. Unilateral initiatives should be continued over a considerable period regardless of the degree or even absence of reciprocation.
10. Unilateral initiatives must be as unambiguous and as susceptible to verification as possible.[31]

These rules were not like those presented in chapter 1: these were to be followed with judgment and care, not algorithmically. Machines could not perform GRIT precisely because it was adapted to how our minds worked psychologically—which was why it might work. The audience for these rules was heterogeneous: the rules to maintain security were directed domestically, to persuade politicians that GRIT was safe; and the rules for reciprocation were intended to demonstrate sincerity to hostile agents, and so were directed outward. All of this presumed that GRIT leveraged actual mechanisms of the human mind.

3.3. The Arms Race in a Terrarium

Why did Osgood think this would ever work? In his initial article, and then again in his *An Alternative to War or Surrender* (1962), Osgood invoked a finding from perceptual psychology.

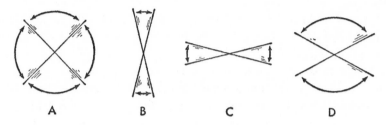

Figure 3.4. Light-bar experiment used by Osgood to explain the principle of incongruity. (Charles E. Osgood, "Reciprocal Initiatives," in *The Liberal Papers*, ed. James Roosevelt [Garden City, NY: Anchor Books, 1962], 177.)

The experiment involves two bars with lights on their ends (figure 3.4), situated at first (in image A) at right angles to each other. The lights on each bar blink in turn. In a dark room, some subjects will perceive the setup as a single bar oscillating around the vertical axis, and some will perceive it as a single bar oscillating around the horizontal. In B, almost all subjects see the oscillation around the vertical, and in C around the horizontal. The interesting result is in D: when the angle in setup B is slowly increased to reach the level of D, subjects will *still* insist the oscillation is around the vertical, even though objectively it resembles C more.[32] People tend to hold to preconceived views even when circumstances change dramatically. This is why the announcements in the GRIT protocol matter; without them, the conciliatory moves would not be seen as deliberately conciliatory, and the opponent would still perceive the situation as tension increasing. By creating a conceptual contradiction—the conciliatory behavior is at odds with the *perceived* nature of the opponent—the perceiving nation would either have to change their conception of the opponent or radically restructure their internal values.

The second reason GRIT should work built on the first: "mirror imaging." According to this notion, a pair of mutual opponents maintain almost identical perceptions of their counterparts, but with the polarities reversed. That is, American citizens believe that they are peaceful, that their leaders are good, that the Soviet leaders are expansionist, but most Soviet citizens are sympathetic and misled by their government. And their Soviet counterparts believe the same about the Americans. The mirror image sets up the expectations that allow the disorientation to do its work; if the Soviet state did not respond to conciliatory moves by the Americans, Soviet citizens would have to believe that their own state was in fact aggressive, which was internally inconsistent with the rest of the cultural picture.[33] The mirror-image theory, however, was based on anecdotes, with no hard evidence available from polling or other sources.

So how could you directly "test" the plausibility of GRIT, or any other arms control scenario? In the 1960s, there was one leading method: the iterated prisoner's dilemma (chapter 5). For a significant subset of psychologists, these experiments were crucial to modeling the arms race: both sides would be better off disarming, but neither side wanted to have their disarmament exploited by their opponents. Once the prisoner's dilemma was operationalized in the laboratory, one could strip away enough of the mind so that rationality could be laid bare.

Marc Pilisuk and Paul Skolnick began one such study arguing that real-world experimentation was simply too messy to produce reliable, robust results: "But the international arena provides too little control for a test of a theory which, in the more general statement, purports to speak to the underlying process of conflict. The laboratory provides opportunity for greater control. . . . The Prisoner's Dilemma game meets these criteria and provides a most appropriate setting for control."[34] The quarry they were after was the validity of GRIT, a procedure that would seem, with its announcements and its flexibility, to be difficult to reproduce within this highly constrained format. Yet their experimental setup was quite elegant, and is represented graphically in a large matrix (figure 3.5), which provided differential payoffs for arming and disarming.

To capture the back-and-forth features of GRIT, the experiment had numerous bells and whistles, but at its core it was a classic setup. A pair of subjects (from a total of 112 male students at the University of Michigan and Purdue University) were seated across a table separated by a partition. Each subject had a board before him with 5 levers on it, each of which showed a missile on the left, and a factory on the right. Subjects would begin each trial with five missiles showing, and at each move they could convert 0, 1, or 2 of the levers to factories or back to missiles. They would announce any intentions to be conciliatory at the beginning of each trial, and there were 5 moves in each trial in which they could either make con-

NUMBER OF MISSILES OTHER PERSON HAS LEFT

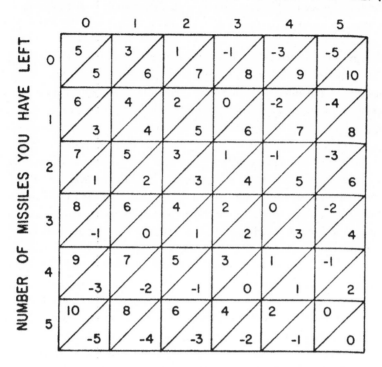

Figure 3.5. Grid used to calculate payoffs in the 1968 experimental verification of GRIT. (Marc Pilisuk and Paul Skolnick, "Inducing Trust: A Test of the Osgood Proposal," *Journal of Personality and Social Psychology* 8 [1968]: 124.)

ciliatory or aggressive choices. At the end of the fifth move a light told each player the state of the other side, and each side could calculate their payoff according to the above matrix. The game was played for 25 repeated trials (125 total rounds), although the students were not told in advance how long the game would last. The principle of payoffs (in this case, monetary) was to reward a subject for his conversion of missiles to factories, to punish him doubly for the degree he disarmed in excess of his opponent, and to reward him doubly for his missile superiority. "The central prediction was that honest inspection and conciliatory behavior would increase cooperation (disarmament), that is, that the Osgood proposal would work," and that prediction was fulfilled.[35]

The other dominant way to test theories like GRIT—and one perhaps more suited to the textured picture of mind Osgood advanced—was to search for historical exemplars in which the GRIT technique (conciliatory announcement, tension-reducing actions, punishing exploitation and rewarding reciprocation) was undertaken either wittingly or unwittingly by real historical actors. The most commonly cited case was the "Kennedy Experiment," when John F. Kennedy announced at a speech at American University on June 10, 1963, that he was unilaterally stopping all American atmospheric nuclear tests, and then Soviet leader Nikita Khrushchev responded by halting production of strategic bombers, followed by more American moves. The experiment, if it was one, had its roots in the Cuba–Turkey missile trade of October 1962 and was terminated by Kennedy's death.[36] Kennedy might indeed really have actually been trying to implement GRIT, Osgood later claimed, since the psychologist had extensive contacts with John McNaughton, Kennedy's assistant secretary of defense for international security affairs, who told Osgood that the president had read the latter's essay in *The Liberal Papers* and implemented GRIT conciliations on local scales in Berlin and—interestingly—during the Cuban Missile Crisis.[37]

Other instances followed. Osgood suspected that Leonid Brezhnev attempted a GRIT-like announcement in October 1979 (inspired by Osgood's presentation at the 1977 Pugwash Conference in Moscow), but that President Jimmy Carter, following the advice of his national security advisor, Zbigniew Brzezinski, stalled the reciprocation cycle.[38] Empirical examples continued to be proposed: test-ban negotiations, tension reduction on the Korean Peninsula, environmental modification warfare treaties, Nixon going to China, the Camp David negotiations between Israel and Egypt, and Willy Brandt's *Ostpolitik*.[39] The most common post-Kennedy invocations of GRIT centered around Mikhail Gorbachev and the end of the Soviet Union.[40] The validity of these examples remain strongly contested by academic and governmental experts on arms control.

Both strategies—experimental simulations and empirical examples from the Cold War—were ostensibly attempts by advocates of arms control to move nuclear strategy away from the escalation-prone models of Herman Kahn and his comrades-in-arms. But these strategies also required that the internal logic of GRIT held true, which in turn demanded confrontation with the model of rationality implied by nuclear strategists. Osgood's critique of this vision of rationality was never stated directly as such, but an analysis of his various invocations of the term illustrates how he embedded rationality within the reasoning subject—and thus the subject was irreducibly prey to the vagaries of reasoning by virtue of simply possessing a human mind. He considered himself gripped by "an intense devotion to rationality,"[41] but took this to imply *more* recognition of our rootedness in our psychology, not less: "Yet the only way any of us can escape constraints on rationality is by recognizing them in ourselves as well."[42] For Osgood, self-conscious choice was a crucial element of rationality; one was not simply rational by default, for without recognition of the emotional and "Neanderthal" roots of human psychology, one's attempts to be algorithmically rational would produce manifestly "irrational" outcomes. This was quite different from Herman Kahn, who also believed in symmetry between opponents, but insisted that both sides would naturally, out of sheer self-interest, produce decisions through algorithmic rationality. For Osgood, however, symmetry meant not that all interactions could be reduced to 2×2 matrices, but that the same constraints on full-bodied reasoning applied to all members of the species. Formalized rationality was simply empirically inadequate; it did not capture experience:

> It is true that men can be rational, and often are, but it is also true that they can be nonrational, and these are merely some of the mechanisms of nonrationality. These are some of the ways in which humans reach decisions without the benefit of logic and without even maximizing their own self-interest in the game-theoretic sense, and yet these ways of thinking are lawful in that they conform to and are predictable from the principles of human behavior.[43]

Nonrationality is not quite the same thing as irrationality, and Osgood's loose use of the language allowed him to slide between the two often without comment. Fortunately, Osgood had developed a way of determining whether one's opponent was rational: GRIT! By building a strategy for tension reduction on his own conception of rationality, he had a test of whether adversaries were themselves rational. As he testified before the US Senate Committee on Foreign Relations on May 25, 1966, "This strategy does assume some rationality on the part of the components; it probably would not work against another Hitler. But by virtue of its carefully graduated nature, it allows us to find out if we are in fact dealing with a rational opponent."[44] If the opponent does not respond to GRIT, one may assume their nonrationality (even irrationality), and deploy different tactics, ones less likely to promote the kind of de-escalation seen in Cuba in 1962.

3.4. Irving Janis on Playing Well with Others

The goal for Osgood, and for many who thought like him, was to understand how our emotions interfered with what was "obviously" the correct choice of de-escalation. One classic problem from the Cuban Missile Crisis, to pick the canonical example, articulated long after the fact by Robert S. McNamara, Kennedy's secretary of defense—perhaps the man most responsible for elevating the position of technocratic rationality within the US government in the 1960s, and one of the most visible public figures associated with the frame of reasoning described in this book—was fear:

> Now we're getting down to an absolutely fundamental point. There was just too damned much fear in the missile crisis. If you keep piling it on, people may crack. You do not want to reduce your leaders to quivering, panicky, irrational people, do you? Look, we've been piling it on ever since the missile crisis, and even before. That's my point: piling up weapons, reducing flight times, creeping toward a functional launch-on-warning posture—all of these were much less troubling in the missile crisis than now [1989] and there was, I assure you, plenty of fear to go around in October 1962.[45]

McNamara and Osgood had common ground here: the foibles of the individual human mind were responsible for distortions of rationality. If you wanted good decisions, you needed either to tame that fear or build a mechanism that would channel it.

The earliest journalistic descriptions of the Cuban Missile Crisis presented its peaceful resolution as the product of calm rational calculation by sober minds, like President Kennedy's.[46] Ever since the posthumous publication in 1968 of *Thirteen Days*, Robert Kennedy's memoir of the events of October 1962, accounts have increasingly emphasized tempers, egos, and other nonrational factors during the ExComm deliberations. Kennedy was careful to introduce the theme of emotionality gently:

> It is no reflection on them that none was consistent in his opinion from the very beginning to the very end. That kind of open, unfettered mind was essential. For some there were only small changes, perhaps varieties of a single idea. For others there were continuous changes of opinion each day; some, because of the pressure of events, even appeared to lose their judgment and stability.[47]

This was no surprise, for "that kind of pressure does strange things to a human being, even to brilliant, self-confident, mature, experienced men."[48] On examining the (secret) verbatim recordings of the ExComm meetings, one might even conclude that Kennedy *overemphasized* this feature. Those tapes demonstrate many fluctuations in emotion—laughter, tension, and so forth—but they are dominated by a businesslike tone and demonstrate that stress and fatigue were marginal to the decision-making process.[49] But amidst this attention to individual emotion, very little was said throughout the 1960s about the psychological implications of the most salient feature of the ExComm: that it consisted of a defined set of individuals. In 1972 Irving Janis coined the term "groupthink" to describe how the relevant psychological unit for foreign-policy affairs was the deliberative *group*. Although he and Osgood disagreed on many aspects of rationality, both concurred that, left unmodified, either the individual or the group would fall prey to psychological forces that could provoke military and political disaster. Groupthink, then, ostensibly represents another effort in questioning rationality's ubiquity in foreign policy; but, as we shall see, Janis's thesis in *Victims of Groupthink* marked nothing so much as the tail end of the rationalist stranglehold on nuclear strategy. Janis's point was not simply that we failed to be rational subjects in the sense of deliberating according to logical rules, but rather that we failed in this because we had taken insufficient account of psychological distortions. Essentially, he defined the irrational as simply the obverse of axiomatized, algorithmic rationality, without interrogating the accuracy of the latter's picture of the mind (as Osgood had). This positioning of psychology as the domain of the "irrational" served, more subtly, to reinscribe the core notion of rationality that undergirded the framework of the hawks on the Right.

Groupthink has taken on a life of its own in popular discourse, but it started out as a fairly narrow concept. Building on two decades' work on "social cohesion," Janis applied the findings of group dynamics to historical exemplars of American foreign policy making and reached a profoundly counterintuitive conclusion: rather than being more judicious processors of information, groups possessed a marked tendency to produce biased decisions *as a function of their structure.*[50] *Victims of Groupthink* explored six historical exemplars, four demonstrating the negative tendencies of collective decision making and two showing the partial taming of these forces. The four negative exemplars were the Bay of Pigs (1961), escalation of the Korean War (1950), the failure to predict the Japanese attack on Pearl Harbor (1941), and the escalation of the war in Vietnam (1964). (In a second edition, Janis added a detailed study of the Watergate cover-up [1973] as a non–foreign-policy example.)[51] Reasoning inductively from a limited source base—mostly published historical accounts, memoirs, and journalism—Janis catalogued six deviations from rationality:

First, the group's discussions are limited to a few alternative courses of action (often only two) without a survey of the full range of alternatives. Second, the group fails to reexamine the course of action initially preferred by the majority of members from the standpoint of nonobvious risks and draw-backs that had not been considered when it was originally evaluated. Third, the members neglect courses of action initially evaluated as unsatisfactory by the majority of the group. . . . Fourth, members make little or no attempt to obtain information from experts who can supply sound estimates of losses and gains to be expected from alternative courses of actions. Fifth, selective bias is shown in the way the group reacts to factual information and relevant judgments from experts, the mass media, and outside critics. . . . Sixth, the members spend little time deliberating about how the chosen policy might be hindered by bureaucratic inertia, sabotaged by political opponents, or temporarily derailed by the common accidents that happen to the best of well-laid plans.[52]

Janis selected the term "groupthink" for its Orwellian feel and thus its "invidious connotation": "Groupthink refers to a deterioration of mental efficiency, reality testing, and moral judgment that results from in-group pressures."[53] The book was a stunning success, especially among the for-eign-policy community.

This was partly because not all decisions were fiascoes, for Janis explored two cases in which the seemingly inevitable tendency of groups to push toward premature and myopic consensus-formation was averted, largely because of how the decision process was structured: the Marshall Plan and the Cuban Missile Crisis. The touchstone source for the "emotional" Cu-ban Missile Crisis was Robert Kennedy, who pointed out that "to keep the discussions from being inhibited and because he did not want to arouse at-tention, [the president] decided not to attend all the meetings of our com-mittee. This was wise. Personalities change when the President is present, and frequently even strong men make recommendations on the basis of what they believe the President wishes to hear."[54] Janis cited this section of *Thirteen Days* (among others) to emphasize that President Kennedy had ensured representation of all points of view, and the deleterious failings of group dynamics did not manifest themselves. Janis's explanation for Ken-nedy's perspicacity was that the debacle of the Bay of Pigs (a canonical in-stance of groupthink) forced the president to reengineer the deliberation process. Even the danger of the situation itself might have been responsible for the happy outcome:

Perhaps the magnitude of the obvious threat of nuclear war was a major factor that, along with the improved decision-making procedures used by the Executive Committee, operated to prevent groupthink. It seems probable that if groupthink tendencies had become dominant, the group would have chosen a much more militaristic course of action and would have put it into operation in a much more provocative way, perhaps plunging the two superpowers over the brink.[55]

So began the amazing career of a concept, yet it was mostly deployed as a vague generalization, a catchall to describe what happened every time a group made a bad decision, whether or not it adhered to Janis's quite specific formulation of the antecedents of groupthink. Psychologists soon demanded more explicit studies to determine not just whether all of Janis's criteria were necessary and sufficient, but even whether the phenomenon occurred at all.[56] After all, a group could make a bad decision without suffering from myopia about alternatives, the emergence of self-appointed "mindguards" to shut down discussion, or overvaluation of the leader's wisdom. In response, a broad array of laboratory tests probed specific components of Janis's model.[57]

The testing at first seemed to lean in favor of confirmation, but by the early 1990s, in the wake of prospect theory and other reformulations of rationality (which we will explore in chapter 6), the only consensus about groupthink seemed to be that the model was misleading and needed to be replaced. If cohesive groups flattened their decision processes and reflexively cohered around risky decisions, then how was one to explain the many instances in which groups did and do make decisions that are rational, by whichever definition you chose? Since Janis quite clearly opposed military intervention, he selected as "fiascoes" those cases where such intervention went poorly, and his "good" cases featured de-escalation. As a result, he—and the groupthink advocates who followed him—cherry-picked cases where memoirs indicated precisely the factors that Janis believed he had uncovered inductively.[58] In addition, there were real efficiency losses with "antigroupthink" procedures; the institutional costs would often outweigh the substantive problem-solving benefits.[59] But the notion of groupthink—its protean definition, its facility of explanation, and perhaps also the way it rolls off the tongue—proved so seductive that despite the empirical problems it became the *only* phenomenon in group dynamics that was discussed in introductory psychology texts.[60]

Groupthink has lost much of its currency for psychology, but it retains its charm for political science. In that, it has outlived GRIT. But even though Janis seems on the surface (like Osgood) to critique Kahn's vision of rational strategy, this brief examination suggests it was more of a complement than an attack. Groupthink is not so much a theory about how we make decisions, as about how decision making can be distorted by psychological drives, with the implication that we could engineer a better outcome. If we heeded Janis and followed certain procedures—like the ones Kennedy followed during the Cuban Missile Crisis—we would tame our natural tendency to groupthink and expose the problem to rational resolution.

3.5. The Taming of Cognitive Dissonance

The most important fact about the Cuban Missile Crisis is that it ended without nuclear war—this is its mystery. Nuclear strategy embraced with vigor the ideal type of Cold War rationality—lean, even emaciated, stripped of emotion and desire, calculating utilities in an environment of uncertainty—and yet its central data point, the one moment when the world had been closest to the brink of a full nuclear exchange, eluded analysis through a rational-choice framework. One might have expected Kahn's theories (regardless of what Kahn himself would have advocated) to predict some aggressive response—say, a limited air strike—to indicate America's willingness to go to war, which risked provoking Soviet retaliation, escalating into the thermonuclear destruction of civilization. Yet the Americans did not strike Cuba but opted for a blockade, and the Soviets backed down, both outcomes (in some sense) not "rational." How could one explain the absence of catastrophe?

For Charles Osgood, the reason the Cuban Missile Crisis was resolved peacefully had nothing to do with strategy and everything to do with tactics: overtures of conciliation, properly announced and followed-through, generated a cycle of de-escalation. A rich picture of the mind that included emotions, biases, and other nonalgorithmic features led to an account of "rationality" that was far from mechanical; in fact, the very rootedness of the mind in our human phylogeny and ontogeny enabled just the sort of tension-reducing outcomes as happened in 1962. The fact that Kennedy had read a version of Osgood's theory cemented the latter's conviction that he had uncovered a universal strategy that would harness our internal motivations to rein in Herman Kahn's escalatory ladders. Osgood knew this would work because he had incorporated something into his theory that Kahn, through his ignorance of psychology and his focus on procedural calculation, had excluded.

The roots of Osgood's insight came from his development in 1955 of his "principle of congruity."[61] That is, when subjects were confronted with clusters of meaning that apparently violated coherent principles of ordering, the subjects would rearrange either their environment or their internal concepts to minimize the incongruity. This principle was later developed by Leon Festinger who gave it the popular moniker of "cognitive dissonance"—with due citation to Osgood's prior work.[62] Festinger's theory grew out of both a Ford Foundation grant to catalog psychological concepts and an ethnographic study of a prophetic movement, but it soon became part of an (ultimately successful) effort to reformulate the learning theory of Clark Hull, Osgood's dissertation advisor.[63] Cognitive dissonance (among *nations* as well as people) would play a major role in GRIT: by manipulating others' expectations of what you were supposed to *do*, you could in turn nudge them toward reevaluating who you *were*, as their minds readjusted to bring the empirical phenomena ("he is making conciliatory gestures") in line with their assumptions ("he is naturally aggressive"). Enough contradictions posed by the former would nudge the latter into a new groove. Cognitive dissonance became a pillar of psychology as it emerged from a cocoon of behaviorism in the so-called cognitive revolution of the 1960s. Among its many applications were occasional attempts to explain the arms race as either the result of or inflected by cognitive dissonance, ranging from the construction of shelters to, yes, even the decision by Khrushchev to place missiles in Cuba.[64]

Cognitive dissonance found its place in Janis's groupthink as well, but in the opposite sense. While for Osgood cognitive dissonance was a mechanism to reduce tension, for Janis dissonance was a disease that afflicted groups of decision makers. Since members of a group "cohered" and believed that they were reasoning properly, the very force of cognitive dissonance pushed them toward groupthink—the assumptions were always reinforced by the empirical evidence of *others'* commitment to them. The way to tame groupthink, therefore, was to reduce those pressures that accentuated the dissonance: self-valorization of the group, minimal exposure to external ideas, and so forth. Janis's vision of rationality was thus closer to the stripped-down, RAND version—subjects were in fact capable of reasoning in a cool calculated manner—provided one took into account the unconscious distortions the environment could pose: psychology could account for the "mistakes" in rationality, once those mistakes were defined as deviations from an algorithmic framework.[65] For Osgood, rationality was a *choice*, and cognitive dissonance was one of the "constraints" that we could harness to achieve it; for Kahn and Janis, rationality in the context of the

Cold War was just what we were naturally capable of doing (although Janis insisted that we didn't do it as often as we might believe).

As the case of nuclear strategy makes clear, there was no straightforward way of adding the mind back into questions of rationality. The appeal of the algorithmic definition of rationality was precisely that it avoided the messes that ensued when one attempted to account for the fact that decision makers had personalities, histories, and prejudices—it would be simpler, perhaps even *better*, just to ignore those as irrelevant to rational choices. Voices like Osgood's were rare, and they were also problematic. In order to make a credible claim against mechanical rationality, one needed to be able to demonstrate that one's claims were reliable, and GRIT did not lend itself to the dominant modes of testing (neither, for that matter, did groupthink). The question of proper method in the social sciences was in many cases even more pressing than the formulation of theories or tactics of behavior, for method was the ground on which the debates over Cold War rationality were waged.

FOUR

"The Situation" in the Cold War Behavioral Sciences

During the topsy-turvy early years of the Cold War, certain places once deemed irrelevant and remote attained sudden relevance. Take Micronesia, for example. At one time, some joker, in a "take-my-wife" spirit, might have tacked on the riposte, *Please*. Ranging across one million miles of the northwest Pacific, the Micronesian territory was made up of three island groupings—the Marshalls, Carolines, and Marianas—that comprised a total landmass of just over seven hundred square kilometers, an area only four times the size of Washington DC. The often hilly, beach-dominated, soil-poor, and mineral-poor atolls had little to offer an outside force aside from their apparent need to be properly occupied and administered. For a long time no colonial or neocolonial power wanted these small islands very badly. In 1919 with the Treaty of Versailles, the islands became a not-fiercely-contested prize taken from Germany and settled on Japan. The Japanese exploited some copra (made from the plentiful supply of coconuts), founded a few bauxite mines, and set up telegraph lines in the more travelled spots, but otherwise neglected the islands.

That changed with World War II, of course. Serving as some of the "stepping stones" in Admiral Nimitz's naval progress across the Pacific to Japan, Micronesian atolls such as Palau—known ever after by American troops as the site of the legendary Peleliu, or "hell in the Pacific"—hosted some of the bloodiest battles of that campaign.[1] By the end of the war the Micronesian islands, first as an Occupied Area and next as a Trust Territory, were no longer to be taken lightly. In the peacetime that followed, they found themselves the site not of concentrated killing but concentrated scientific inquiry. How far could social scientific research penetrate into the rational and irrational parts of individuals' personalities and cultures? Under the banner of scientific inquiry, varieties of madness and breakdowns of rational behavior came under study in a circumscribed area.

It was not that any economic value had manifested itself. Rather, their remote location, their "tiny world" quality, the way they sat as bits of land surrounded by "a million miles of ocean," enabled—demanded, apparently—the pursuit of several kinds of research. For American science they offered an unparalleled opportunity. To put it bluntly, the islands could serve as intensive nuclear test sites for the physical sciences and intensive psycho-social test sites for the behavioral sciences. They constituted a unique situation, a de facto offshore laboratory where a new form of Cold War rationality could be test driven. Research not possible elsewhere was possible there. Soon to be known colloquially as the "Nuclear Pacific" and officially as the Pacific Proving Ground, the islands' scientific status outsized their diminutive stature. As Admiral Chester W. Nimitz wrote to one social scientist working there, "science itself," in addition to the US Navy, was the beneficiary of these coordinated studies.[2]

In this way, the Micronesian atolls were a curious combination of far-off and central, of irrelevant and momentous.[3] Their adoption as test situations in the social and physical sciences intensified this paradox. In the most basic sense, the paradox appeared in the term "Trust Territory" when its definition was extended to indicate not only a place protected by the paternalistic presence of the United States but also as a place for free experimentation and a degree of intensive testing not possible at home. (On certain atolls, as we will see below, nearly 100 percent of the population received in-depth psychological and "moral" testing.) In this sense it posed a paradox typical of Cold War rationality: although it sprang up in direct response to geopolitical and administrative emergencies, it encouraged a form of thinking and practice that was curiously abstract and obsessed with method over gritty substance. A real place became partially unreal, in the sense that it gained laboratory-like status. As with other examples of Cold War rationality in its project-based forms—RAND's prisoner's dilemma games tested in living rooms, Berlin's Tempelhof Airport turned into a site of algorithmic programming, the use of a gypsum factory to spell out rule-based human behaviors—it drew people to collaborate and fertilize across disciplines.

Spots like Micronesia were not merely test sites, they were sites for testing the unprecedented use of intensive testing. Led by Yale's George Peter Murdock, the Coordinated Investigation of Micronesian Anthropology was designed to be more complete than any previous study of a single place.[4] Forty-one social and behavioral scientists from over twenty differ-ent institutions fanned out to investigate in depth. In addition to covering straightforward topics such as an alarming trend in depopulation, linger-ing effects of starvation during the recent war, native use of plants, and "In-sects: harmful and beneficial," researchers were to collect information on islanders' political beliefs—who allied with whom, where sympathies lay. And perhaps of most relevance to our inquiry, psycho-sociological testing too went forward at an intense pace. The question was not so much how profoundly could you understand a given cultural group but how *quickly* and efficiently could you amass results via standardized methods. (Recall that an emphasis on speediness of calculation was also seen in chapters 1 and 2.) Ethnographer William Lessa gave Rorschach tests to each resident of Ulithi, along with specially designed thematic apperception tests pictur-ing tropical themes of gales, canoes, grass skirts. Nearby on Truk, anthro-pologist Thomas Gladwin took life histories for twelve males and eleven females, and gave these documents, along with full TAT and Rorschach re-cords, to Chicago psychiatrist Seymour B. Sarason to analyze. On neigh-boring atolls, Alice Joseph and Melford Spiro administered the tests to a full population sample of Chamorros and Ifalukans, respectively. (Spiro added to the battery the Bavelas Moral-Ideological test as well.) Speed was of the essence—since the goal was optimally efficient research conducted by means of near-automatic methods—and it often seems (to the onlooker today) as if researchers cared as much about start-to-finish time clocked as the richness of their results. In-depth local knowledge was not an aim, per se. It was typical to note, as Joseph did, that their report was not meant to be a complete, penetrating study, but that "the work was planned experi-mentally as a deliberate attempt to find out how much information con-cerning personality structure in a cultural group can be obtained by a rela-tively short, standardized method in cases where there is not enough time for detailed personal and individual studies or systematic research on social and cultural backgrounds."[5] In the coming decade, these researchers passed their "raw" results on to collective encyclopedic enterprises such as the proto-database "Primary Records in Personality and Culture."[6] To test the limits of testing within a circumscribed area in a limited time was the aim.

Somewhat monstrously, the context of this welter of tests was targeted nuclear Armageddon. While the human experiment proceeded on cer-tain islands, physical experiments ran on others in the same area. Sixty-six atomic bombs exploded in Micronesian territory over twelve years, acceler-ating in frequency, intensifying in force. Beginning in the summer of 1946, the Joint Chiefs of Staff's Operation Crossroads took Bikini atoll and surrounding waters, in the thick of the Coordinated Investigation's range, as its target for two successive nuclear tests hitherto unequalled in tonnage. In the ensuing years from 1946 to 1962, even as the islands received 14 percent of all US nuclear tests—and absorbed over 80 percent of the total yield of nuclear bombs detonated in that time—they continued to undergo intensive study by physical anthropologists, ethnographers, linguists, sociologists, and "human and economic geographers," among others. No less than 210 megatons dropped in the vicinity, but, undaunted, social scientists pursued their own intensive tests (figure 4.1).

The strangest part of all: neither enterprise to our knowledge *ever referred to the other*. A curious abstraction in attitude meant that the sociopsychological experts on the one hand and the nuclear on the other operated as if their "experimental zones" did not share the same geography. Their respective results bore more on the sciences to which they were meant to contribute than the living context they inhabited. Whole islands vaporized, even as new worlds of social and psychical information came into being. The landmass of Elugelab may have disappeared one November day in 1952, but its information was preserved. As a situation, Micronesia was multiform and improvisational, the host to all kinds of tests and experimental methods (figures 4.2 and 4.3).

Figure 4.1. "Baker Shot," part of Operation Crossroads, took place at Bikini Atoll in 1946. Nearby, in 1947, a group of thirty-one American social scientists began the Coordinated Investigation of Micronesian Anthropology, a thorough multi-atoll study of Micronesian psychology, culture, and physiognomy. The project never mentioned the escalating series of US nuclear tests in the area. (Reproduced courtesy of U.S. Department of Defense.)

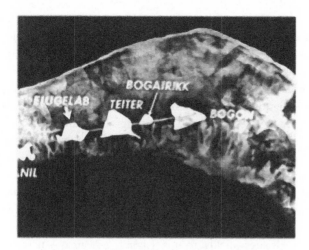

Figure 4.2. Micronesian atolls before Ivy Mike shot
(hydrogen bomb) of November I, 1952.

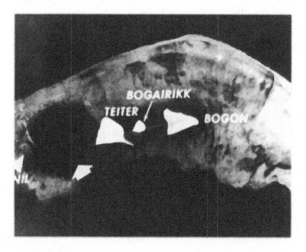

Figure 4.3. After: A crater was all that remained of the island of
Elugelab, Enewetak Atoll, Marshall Islands, US Trust Territory.
(Reproduced courtesy of U.S. Department of Defense.)

4.1. Into the Microcosm

Micronesia's dollhouse opportunities—where researchers multiplied experimental tasks, gave tests of all kinds, and outfitted methodologies for operational use—were not unique. (Even among anthropologists and psychologists, the Micronesian project was only one of many designated "field laboratories" in these years that harnessed techniques from across neighboring disciplines.[7]) With claim-staking methodological flair, across rapidly unifying fields and subfields in sociology, cultural anthropology, social psychology, economics, early cybernetics, and operations research, among others, specialists marked out a set of laboratory-like scenarios and quasi-laboratory situations. Designated domains included "special rooms" rigged with recording devices and secret adjoining observational facilities where scenarios could be played out under close study; occupied areas under US military control stipulated as "experimental areas" for tests alternately nuclear and psychological; field sites labeled substitute labs or offshore investigatory spots; and several kinds of "strange situations," test situations, and coercive situations for close study of the forms and functions basic human relationships might take under different conditions. Boosters promoted the use of such sites—fields, rooms, islands, test areas—in methodological manifestos, ardent if somewhat drily expressed calls to arms. All were something like laboratories and something like unadumbrated life, but did not count fully as either.

Common to many such sites was a term of art—the "situation"—that emerged as a multifarious and nearly all-purpose construct among groups of allied and cross-disciplinary social scientists. At its most basic level, the situation was a mechanism to create consistency in the phenomena being studied. Complex as they may have been, they could be contained within a designated situation and thereby understood. Consistent, comparable units of behavior were in this way laid out for study. As an additional effect, the situation would create possibilities to exert control over its constituent units. Thus the situation was a methodologically defined space for exploring how Cold War rationality, defined as an ongoing argument about the best way to operationalize rational features of human social life within a constituted system, could be found in something *resembling* real life and (potentially) made to operate in other situations.

The situation arose in a corpus of methodological thought common to American social scientific innovators. A postwar "behavioral revolution" brought together quora of empirical, quantitative social researchers who shared, as one participant recalled it, "the aspiration for an interdisciplinary, unified behavioral science in contrast with traditional speculative 'grand theories.'"[8] Prewar rat-in-maze-generated behaviorism found new life within the integrative social sciences, in turn newly located under the funding umbrella-category of the behavioral sciences. Here the situation grew in importance. Rats no longer needed to run through mazes. The situation was much more multiform and adaptable. It spread in part via the postwar flourishing of "methodological writing."[9] So commonly was the situation called upon—expressed in formal and informal methodological claims, extensively rehashed or simply ubiquitously if casually employed—that it perhaps qualifies as a "style of thought" in Ludwig Fleck's sense, common to these years and this context.[10]

A tour of the array of definitions and usages of the situation that took root among a variety of practitioners would rightly first include, as background, a turn to John Dewey, for whom the situation was a central concept. During the 1930s, Dewey made the situation the core of pragmatic inquiry: a situation was a place of uncertainty—a problem—that could be resolved into harmony through proper understanding and experiment. The situation designated a container that holds important events and behaviors for study. Dewey believed "the sense of confusion engendered by a situation, rendering it problematic, sparks action that seeks to alter that status quo."[11] The situation was both a site of constraint and of possibility. It was the place where a problem occurred as well as its potential to be worked out. Dewey's situation necessitated future action via experiment. This action-based quality (within something that remained, for Dewey, still rather undefined and unspecified) persisted among researchers to come.

For social scientists in the post–World War II years, the situation became both more technically specific and more methodologically nimble. Writers of widely circulated textbooks and manifestos seized on the situation: Kurt Lewin for one carved out the "total situation" as a heightened space of observation and intervention.[12] Following Lewin (if sometimes reinterpreting his views), social psychologists defined their laboratories as situations, attempting in this way to lend them all the qualities of "real life": a lineage extending from Leon Festinger to Solomon Ash to Elliot Aronson and J. Merill Carlsmith forcefully extended their experimental claims by means of the situation. "When we take the time and trouble to build experimental reality into the procedure, we succeed in creating a situation where, within the confines of the lab, real things are happening to real people."[13] Situations were both real and artificial. Through such methodological audacity, a rising generation emphasized the experimental opportunities situations offered.

Situations offered "elbowroom for action," to borrow a phrase.[14] Within American mid-century sociology, the work of Paul Lazarsfeld at the Bureau of Applied Social Research also privileged the situation as a site of interaction.[15] Perhaps paradoxically, an experimentalist could potentially carry out a more radical manipulation of subjects and relationships in such a specially circumscribed situation, or, alternatively, could make more radical claims for future abilities to alter behavior, because he was less constrained by the built-in limitations of working with "human materials" in formal lab-bound inquiry. As sociologist Robert K. Merton put it in 1946, "a 'real life' situation rather than a synthetically arranged event" when properly approached offered advantages over a strict "'artificial' laboratory."[16]

And to return briefly to the anthropological field situation, a common understanding was of the field as an experimental area, one not simply for participant observation but for going further: "participant intervention," also known as "experimental intervention in the field."[17] Pop-culturally speaking, back-up can be found in the views of *Candid Camera* founder Alan Funt, who influenced social psychologists such as Stanley Milgram. Funt explicitly sought "pure situations" where an onlooker could view "untainted and unadulterated sides of human social behavior" so that mechanisms, behaviors, and rituals of everyday interaction were exposed.[18] (Likewise, the rise of the TV "sit-com" occupied Americans' living rooms while the building of a succession of ever-more-reinforced Situation Rooms occupied the White House basement.) Across the newly emerging American behavioral sciences, sites of Cold War rationality, the situation took root. Note that all of these accounts were neutral with respect to what happens in a *situation*. It was not about content, at least not initially. That was up to the investigator to create or locate. What was important was that this was defined as a privileged, experimental space—one that was not exactly a laboratory and not exactly the real world.

This chapter asks, How did the *situation*—as methodologists defined, re-defined, and deployed it in its various forms—affect the way people thought, and the way they thought about how other people thought and acted? In short, what specific styles of rationality arose in the experts who carefully plotted out and defined such situations? What follows examines closely one of the most deliberate and successful (if often Quixotic) attempts to confine the gamut of human behavior in a methodological rigorous and enclosed situation, the Special Room at Harvard's Laboratory of Social Relations under Robert Freed Bales. New scientific methods helped researchers to attain a "persistent attitude of abstraction" toward the places and cases where human interaction with other human beings is most tested—such as human frailty, distress, stress, bonding, clinging, loving, following, arguing, lapses of rational behavior.

As the polymathic pioneer of studies of rationality and its limits, Herbert Simon, put it in a mid-career lecture, all the social sciences aimed to explain one thing: in what sense you could say, with Aristotle, that man is a rational animal. The "irrational is the boundary of the rational," he scribbled in lecture notes after surveying results in anthropology, sociology, psychology, and economics. "And the boundaries determine behavior." He followed this with the somewhat inscrutable yet double-underscored comment: "bowl of molasses."[19] On second thought, bowl of molasses captures our topic exactly. Were the bowl not there, the molasses would spill out everywhere. This is exactly what the situation did: it kept the molasses in, so one could study and learn what part of it is made of this component, what part made of other ingredients. What made people act one way or another, behave rationally or irrationally, and could all this oozy stuff be held in one bowl? Yes, one could say, if the bowl were a situation, and if its contents were properly analyzed.

4.2. The Special Room as a Situation: What Made it Special

One of the greatest lab-like "situations" in this era lay in the work of Robert "Freed" Bales on the small group. Bales was one of the "young men in their early thirties" whom Chicago-trained sociologist Samuel Stouffer brought in to work in at Harvard's Laboratory of Social Relations, the hands-on "shop" developed to support Talcott Parsons's Department of Social Relations. As an enterprise, Social Relations embodied postwar hopes for unifying what were seen as the key American social sciences. In the immediate postwar climate of optimism and rethinking, the new experiment in disciplinary collaboration got its go-ahead vote from the faculty on January 29, 1946. At root, its instigators wanted Social Relations to mount a vast moral and legitimating structure for social life itself—"a regenerated system of morality," in the words of one of its founders, Henry A. Murray[20]—even while claiming, with all the firepower of fervent scientistic language, that they were making no moral claims at all. In this Harvard was trying to make a social science along the lines of a physical science, one that could explain and predict human behavior much as physics had explained and predicted atomic behavior.[21] The Manhattan Project inspired them in turn to try their hand at "splitting the social atom," as Parsons used to like to say.

The Laboratory's first five-year report described its plans for promising young men like Bales in terms otherwise found in YMCA literature: "Once such a man comes, he is free to work at anything he chooses. There are few big projects, and none with elaborate administrative overhead and division of labor. The researcher can work by himself or he can team up with a congenial colleague. Of course, the Laboratory had a pretty good general idea of his interests and his skills when he was first added to the staff in preference to other men."[22] In this "seedbed" environment, Bales distinguished himself almost immediately and rose from doctoral researcher to assistant professor in only three years, working closely with Talcott Parsons, among others. (He even won tenure in Social Relations, a rare accomplishment.) It is hard to exaggerate the hopes that hung, for a time, on Bales's research in the intricate situation he constructed. Having embarked on his earliest (pre-Harvard) research to answer the question, *"What is meant by the situation?"*—his 1941 master's thesis was titled "The Concept 'Situation' as a Sociological Tool"—Bales for the remainder of his long career expanded and intensified this inquiry.[23] At Harvard's Laboratory he built an experimental space of his own design, the "special room"—which he also called the "observational situation"[24]—to explore exactly how much one could learn in one such a methodologically honed situation. The focus here is on what took place in the space Bales made. Among other things, its proceedings helped to solder the newly formed Department of Social Relations to its Laboratory and served to support a new kind of experimental rational actor.

Encouraging Bales's work "to the best of our ability was one of the early policy decisions made on the establishment of the laboratory," Parsons recalled its head, Samuel Stouffer of the wartime *American Soldier* project, saying.[25] (Note that Parsons was head of the Department and Stouffer head of the Laboratory; the much more junior Bales travelled back and forth between the two entities.) From the outset Bales hoped to combine four fields—social anthropology, social psychology, sociology, and clinical psychology—and give his work a firm empirical ground that was yet ripe for theorizing. Small-group research, although obscure today, was not just a subfield of one of the four fields that made of Social Relations, but was the promise for bringing them all together: it was "peculiarly favorable"—as Parsons eagerly affirmed—for the development of the ultimate grand theory.[26]

Bales was far from the only researcher in America working on small groups, but he was one of the best known innovators in this growing subfield of sociology. (Small-group researchers generally credited Georg Simmel's turn-of-the-century work on groups of two and three as their precursor; Bales also cited Max Weber and Dewey as precursors.) During the early decades of the twentieth century interdisciplinary social scientists had been busy studying small groups of people in factories, union halls, and conference rooms, with the Harvard Business School's experiments at the Hawthorne Works plant of the Western Electric Company in the mid-1920s as a landmark case, called the "first major social science experiment."[27] Events at Hawthorne seemed to show that the experimental conditions themselves— under which each worker felt herself to be the focus of interviewers, to be under observation, to "count" in both senses of the word—brought about the desired results of group cohesion and greater adjustment: in sum, a sort of group mind.

Bales trained in the tradition stemming from these pre–World War II studies, but was far more explicit in elaborating his how-to approach. A methodological strongman, he distinguished himself by building a nexus of apparatuses, recording devices, observational protocols, training sequences, and procedural manuals of striking potency. Bales arranged an artificial setting that functioned something like a lab and something like a conference room or generic semi-casual setting, allowing 'natural' if constrained human interaction to occur within. He was the first to designate a so-called "special room" equipped with microphones suspended from the ceiling, and a set of large one-way mirrors surrounded by acoustic tiles on the walls. An adjoining sound-insulated observation room allowed whatever took place on the other side of the mirrors to be observed, recorded, and discussed without disturbing the proceedings. Again, while none of the specific elements of Bales's research design was perfectly original to him— one-way mirror set-ups for unobserved observation dated to the 1930s, for example, in the prescient experiments run by Saul Rosenzweig—it was the combination of techniques, spaces, and procedures that resulted in something new and struck high-theory entrepreneurs such as Parsons as uniquely promising. Bales's first special room was ready for use in 1946; by 1950 his innovation had caught on and there were at least eight or ten such rooms being built across the country. As *Look* magazine described the set up,

[A] thin man who looks like a bank examiner, peers intently through a 'magic mirror' at a group of graduate students around a conference table. They do not seem to know they are being observed. When they look up at their side of the glass through which the thin man, Professor Robert F. Bales, is examining them, they see only themselves in the one-way mirror. They are unaware too of the strange oblong recorder on whose moving tape Professor Bales is making observations for his *Study of Small Groups*.[28]

This popular report emphasizes something uncanny going on in the thin man's room: there is the "strange oblong recorder," the "magic mirror," and the eerie spectacle of a group performing under heavy technologically-aided surveillance without seeming to realize they are being observed. An untoward quality inhabits the set-up in the special room, one echoed in another popular account. Harvard business school professor Elton Mayo in 1948 described to *Colliers* magazine his research situation in the Hawthorne Works factory's observation room as a "twilight zone where things are never quite what they seem."[29] Impresarios of the situation (reporters or researchers themselves) did not hesitate to play up the strangeness of their enterprise. Common assumptions about how the world worked and how people related to each other might be upended through super-hi-tech research designs.

If in prewar studies experimenters had attempted to make their presence less and less obvious, so as to get closer to "real" conditions, at Bales's laboratory the experimenter was more and more present, for the room was itself a "real" condition, somewhere between artificial and ordinary. Indeed, Bales's very definition of a small group was rigged to the precise way they were to be studied within the special room: small groups were *defined as* "groups from size two up to some number of persons who are still few enough to interact together directly in a face-to-face situation."[30] By this definition, which Bales freely admitted was "arbitrary," the researcher defined the small group as that which was small enough to be brought with relative ease into the confines of his own special room. Like agricultural crops modified and grown to fit the machines that will ultimately harvest them, Bales's objects of study (his small groups) existed as entities best studied via his targeted method.

Also by definition, such groups were made up of "face-to-face" encounters among group members. This meant that "a number of technical aids and methods of observation and testing" could be used where they couldn't have been used previously, either because they hadn't yet been invented, or because a mass group like a crowd was too big and unruly to file willingly into a special room with a one-way mirror, magic or otherwise. With the small group in the laboratory one could use gadgets like sound recorders, polygraphs, adapted electroencephalographs, and IBM computers (as Bales did); one could also use machines to measure such physiological responses as heart beat or galvanic skin response, a slight sweating of the palms induced by high stress (as Bales hoped to do in the future). At the Laboratory shop six "Interaction Recorders" of Bales's invention were built. These worked by feeding a moving paper tape through the device at a constant speed, so that an observer could write scores in sequence on the paper as acts occurred in the special room. At the Laboratory's computing facilities, or at a nearby Hancock insurance office glad to lend its additional IBM machines for service, Bales could subject the resultant paper strips of scores to systematic content analysis and interaction process analysis (as Bales had pioneered them). And these strategies *in toto*, as Bales announced, "are making it possible for the researcher to follow the events he is studying in a microscopic and systematic way to an extent that would only have been dreamed of by a science-fiction writer ten years ago."[31] What science fiction had only fantasized about social science had rendered real. No more was there a need for a "control" group, which now "appears crude indeed," and Bales felt that research could now rise to the level of "systematic observation and measurement of free process" that one could find "for example, in astronomy." The space opened up by Bales's special room allowed a style of observation more systematic to describe a process more free. (That is, the space of the special room was exceptionally controlled and this allowed the interaction taking place there to be less fettered than a traditional experiment might allow.) Both advantages were leveraged, and this bode well for the arrival of that general or basic social science of which many spoke.

There were futuristic elements to Bales's approach, not only in its science-fiction overtones but in its pinpoint minuteness, its microscopic focus. As he wrote to a rival who was studying group dynamics in a more old-fashioned way: "We do exactly the same thing [as you do], but *on a more microscopic level of size and time span*."[32] Bales went incrementally smaller than anyone else, because with his special room as his site he was equipped to go smaller. This atomistic approach meant the scientist could exert more control over events that were of the essence of human functioning, events that, if not amenable to laboratory experiment, could be at least be studied in the special room; and the extra margin of control yielded was central to the project's claims to embody a superior form of rational method—especially when rational was taken to mean (on the scientist's part) orderly analysis, well-thought-out methodology, the incorporation of behavioral error into systematic functioning, the thorough coding of components, and the smoothing out of the unexpected (single act) in favor of the predictable (overall pattern). Bales could record each tiny skeletal, verbal, gestural, and expressive behavior of a group member and end up with a graph of what he called the Profile, Sequence, Phases, and Matrices for each group. For example, Bales in one study recorded the interactions of a series of twenty-two groups—including a five-man chess novice group planning the first move of a seven-move problem, a three-man chess group working on a group projective story, an eight-man academic group planning a thesis, a four-man steering committee planning a Christmas party—and drew out from each small group a complete statistical sketch of their labors: a "topical cycle of operations." Bales also brought therapy groups, husband-wife marriage counseling sessions, mother-child interactions, "preschool gangs"—Bales's perhaps overly alarming term for social groupings of four-year-olds—and psychiatric wards into his experimental room. In the miniature world of the special room, small groups stood in for any and all person-to-person social interactions. It held (literally rather than literarily) a synecdoche of society itself—society in the process of being rendered as code.

Each "event" that took place in the room—when one group member said or did something to another—went into one of 12 categories seen as "logically exhaustive of all possibilities." In a cascade from positive to negative, the 12 categories were: "Shows Solidarity," "Shows Tension Release (jokes, laughs, shows satisfaction)," "Agrees," "Gives Suggestion," "Gives Opinion," "Gives Orientation," "Asks for Orientation," "Asks for Opinion," "Asks for Suggestions," "Disagrees," "Shows Tension," "Shows Antagonism." He gives this example of the method:

> For example, suppose we are observing two persons talking together, identified as 1 and 2. Person 1 says, "What time is it?" The observer decides that this is an act of "Asking for Orientation" Category 7. He writes down the numbers 1–2 in the space following Category 7. Person 2 now says, "I have just twelve o'clock" The observer writes down the numbers 2–1 in the space following Category 6, "Gives Orientation." Person 1 then says, "Thank you," and the observer puts down the numbers 1–2 in the space following Category 1, "Shows Solidarity." . . . Person 2 might say, "I have just twelve o'clock. But my watch has been stopping lately. I don't know whether that is the right time or not." The observer would enter three units for this sequence, one for each sentence. If the person had managed to say this all in one sentence, the observer still would have broken it down into three scores, since there are essentially three items of information, or logical points conveyed.[33]

Trained observers in an adjoining room sat categorizing act-by-act the group's ongoing interactions as they raced by and jotted down scores on the moving paper strip of the Interaction Recorder, averaging between ten and twenty scores per minute for most small groups. It must have been stressful work, and obvious difficulties arose in getting five observers to come up with the same scores (when properly trained scorers did not agree, their

Figure 4.4. Robert Freed Bales employs his interaction recorder at the Laboratory of Social Relations, 1949. On the other side of a one-way mirror sits the group he is observing. (Reproduced courtesy of Harvard University Archives, UAV 605, Box 1, Folder: Bales.)

scores were averaged). The whole interaction was sound-recorded, but one could not tell for sure who was speaking by sound alone and therefore the scoring had to be on the mark in real time, while the group was interacting. The observers themselves were being observed, with a bank of lights telling them whether they were coming up with the same scores (figure 4.4).

Still, in spite of rigorous training, only one in three in an early batch of trainees was able to become a "satisfactory scorer" and withal, "There are probably some persons who cannot be trained in any reasonable time to do the job as other scorers do."[34] Bales would try to screen out scorers who suffered from low verbal skills, low manual skills, or "general emotional factors such as negativism or need for autonomy, as well as idiosyncratic personal frames of reference which prevent the trainee from really understanding or adopting the frame of reference assumed by the method." As it turned out, the best scorer was one who operated on auto-pilot, "like a skilled typist or telegrapher" and who often, as a result, "is not able to give a very coherent account of what went on."[35] Often the experimental-

Figure 1

TYPES OF INTERACTION AND THEIR RELATIVE FREQUENCIES

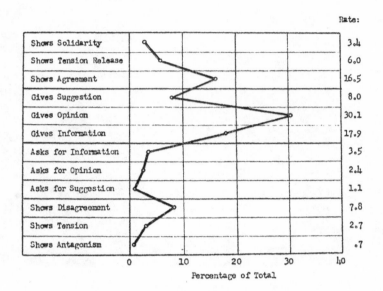

Rate:

Shows Solidarity	3.4
Shows Tension Release	6.0
Shows Agreement	16.5
Gives Suggestion	8.0
Gives Opinion	30.1
Gives Information	17.9
Asks for Information	3.5
Asks for Opinion	2.4
Asks for Suggestion	1.1
Shows Disagreement	7.8
Shows Tension	2.7
Shows Antagonism	.7

Percentage of Total

[This profile of rates is the average obtained from 24 different groups, 4 of each size from size 2 to size 7, each group meeting 4 times, making a total of 96 sessions. The raw number of scores is 71,838]

Figure 4.5. A breakdown of the nine categories of exchange that were performed in a total number of 71,838 social interactions in the Special Room. The typology was intended to be exhaustive of all human interactions that could possibly take place in the room (Bales 1954). (Reproduced with permission of RAND Corporation.)

observer job was one for graduate students and wives of graduate students: pay was low, and job satisfaction not high, bringing to mind the similar job dissatisfaction of the long-suffering lineage of female "human computers" mentioned in chapter 1 (figure 4.5).[36] Having weeded out scoffers, naysayers, and bumpkins, Bales managed to find enough competent and willing "advanced" scorers to secure a .85 correlation among them. This he deemed an adequate degree of reliability and was able by 1950 to proclaim his method overall "theoretically clean and applicable to a very broad range of small groups."[37] By 1955 he declared his Interaction Process Analysis to have been "essentially standardized."[38] He discovered in each small group, no matter what they were up to, a uniform structure: an initial phase of orientation, a middle phase of evaluation, and a final phase of control.[39]

4.3. A Really Ideal Research Strategy

All these high-tech, quasi-sci-fi features made Bales's "the ideal research strategy": ideal research on ideal groups in an ideal setting.[40] A "microscopic method for the study of social systems, of culture, and of personality—all three," Bales called it: the researcher could isolate in a room, around a table, behind a one-way screen, the small group as a microcosm of the world at large, played out face-to-face, writ-small and rendered ever smaller by infinitesimally sensitive recording devices. One could study its workings *at a particular level of molarity;* that is, at a level of *minimum extension* as to space, time, number of persons, and as to the meanings involved in the activity." Like a chemical reaction in its most basic form, the small group within the laboratory showed the most basic workings of those inchoate forces that had confounded social investigators for centuries. A key advantage, too, was that on this level, as in a chemical reaction, the researcher could "introduc[e] changes of the kind he desires" into the group in a controlled manner, through feed-back—information fed back to the group for their own edification and for greater control by the researcher and by the group members of their direction. It was Bales's hope that the irrational (i.e., the implicit or unconscious parts of social interaction) could be made rational (i.e., explicit) through this feedback method: "There is . . . a possibility that if visible indices of implicit emotional processes are made available to the persons engaged in interaction, it will become possible to develop control over these processes through language and other symbolic mechanisms involved in interaction."[41] Out of microscopic measurement would come control—and not just "social control," per se, the common drum roll of the modern social sciences. Control of irrational processes was the aim. Bales's interaction process analysis was the bowl that contained the molasses. Such control would extend to intimate matters of personality and subtle realms of culture.

The more ideal the method, the more real and practical the results. Bales often emphasized that control was of "extreme interest" to his researchers. Not only could each person in the group potentially control his or her interactions, but the social scientist in the observation room might control the entire group, as a conductor an orchestra and each musician his instrument. Bales had found that all groups worked as systems, but some were better systems than others. With his method, he could discover which factors made groups able to reach a "stable and successful" equilibrium state and others unable. These factors, once they were isolated as numerical "profiles," could presumably be extracted and made generally applicable to social life. Through statistics one could derive "empirical norms." These were used for personnel selection, for example, in submarine crews at New London Submarine Base and for "strategic positions in information networks found in operating organizations within the air force."[42] (As we will see, Bales also worked for the RAND Corporation.) Further, his work might apply to committees, that most American of predilections: "Consider, for example, how much time Americans spend in committee meetings of various sorts, and how great would be the saving if the efficiency of such committees could be increased by as little as ten per cent."[43] The point was to isolate in a specially designated and designed room the sometimes messy, sometimes crabby, and sometimes productive stuff of social interactions so that any system could run more smoothly. Should there be "anomalies" or "deviancies," any hiccups or glitches in a group profile, these errors—ultimately invaluable for programming—could be counteracted, smoothed out via feedback, or redirected.

In 1952 Bales announced, "men are now re-combined in such a manner as to provide the best collection of raw material from which . . . a system can be built."[44] Note how unusual is Bales's language here: he asserts nothing less than that his method renders human beings as "raw material" in order to recombine the constituent stuff—coded elements of interactions—into a working system.

Note too that there is no direct search for something one could call "reason," "rationality," or even "quality" within the interaction situation. Content itself was deliberately and explicitly set aside: researchers were not looking for the "best" or most reasonable way to plan a birthday party or run a committee. They were looking to extract the patterning of what possibly could happen, all of human relationships' permutations, foibles, and dynamics. They wanted to be awash in "floods of information available from moment to moment" and synthesize that information via a "dependable device" or a "good instrument."[45] In this Bales's approach was in tune with the experimentalist drive of the 1940s and 1950s, when researchers in diverse fields tried to "adapt . . . the experimental method of physical science to the study of the problems of human relationship," in the words of a 1947 handbook.[46] Face-to-face, what happens, happens. Whatever doesn't serve the progression of phases may eventually be eliminated. Interaction streams by and the trained observer renders the series as code, at a rate of twelve to fifteen codings per minute as a general rule. The fact that it can be so coded and (one day) choreographed is enough. There is no room to judge. There would seem to be no room for reason at all in this system, if we are speaking of Kantian reason, which "must subject itself to critique in all its undertakings," or mindfully deliberate.[47] There is only a kind of two-tiered activity going on: the actors within the room are engaged in a (presumably) rational enough task, planning a party or running a conference or playing chess or strategizing a conflict, occasionally erupting in disorder or dissent. Sealed in a room, they are generating sociological data, making of their room a factory of facts. Meanwhile, the appointed observers are typing away, not thinking at all (recall that thinking on the job is a detriment to doing good work), acting as passive instruments, mere registrars. Just as the simplex algorithm of chapter 2 with its scads of automatized calculation made a virtue of mindless automaticity, so did Bales's interaction method incorporate *not thinking* as a desideratum in its observers.

Still, one can see that an element of rationality is built in to this congeries of rooms, machines, and people, even if it is not always acknowledged. Rationality, Cold War style, is best represented by the figure of method-maker Bales, photographed from behind the back of his head, overlooking the overseers. The silhouetted black scalp and right shoulder are blank, while the front, his face, is not revealed, save for a crescent of spectacle. Here, rationality resides in the relationship of experimenter to situation. More exactly, it resides in the fact that the experimenter, not a personal presence but a methodological overseer, pervades the designated space while his trainees with their prosthetic "eyes" are mechanically coding away on the recording machines they operate and his actors in their activity room are working away on their own puzzle or plan. The space itself is dollhouse-like, miniature, a world apart and yet connected intimately to our own (figure 4.6). It becomes the site of a mini-society creating itself at every moment, or as Bales calls it, a "microscopic prototype of the processes and problems that characterize a wide variety of communications and control networks, both human and electronic."[48]

4.4. Situations for the Situation: RAND

Next for Bales came a second flowering of his work, this time at the RAND Corporation (a player, at least obliquely, in almost all this book's episodes narrating Cold War rationality). Around this time, in 1954, the Balesian method, already quite rigorously standardized, was standardized a bit more. A central accomplishment, Bales reported, was the establishment of a "standard task" that all groups would be asked to complete within the special room—now designated a laboratory, interestingly. Between 1952

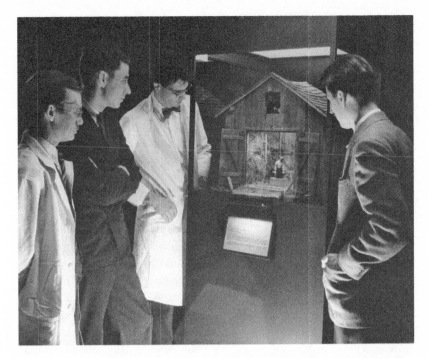

Figures 4.6 and 4.7. Dollhouse tableaux, or "nutshell studies," used in crime scene investigation at Harvard's Department of Legal Medicine. The lab was founded in the 1930s by forensic miniaturist Frances Glessner Lee, a millionaire heiress who loved solving hard case murders in tiny size and building models where difficult or confusing crimes could be explained. *Above,* a group of Harvard medical students closely scrutinizes a forensic diorama: "The Hanging Farmer." *Next page,* an ersatz crime scene constructed by Lee's methods at Harvard's forensic lab, as pictured in *Life*, May 1946. (Reproduced with permission of Getty Images.)

and 1954, Bales worked in Santa Monica on and off, drawing out lessons and lending his work to cybernetic conclusions. He collaborated with John Kennedy of the Systems Research Laboratory, also at RAND, to complete studies for the Air Force "on matters affecting the national security" (described below).[49] Here, the always embedded implications for Cold War rationality, latent in his earlier work, came to the fore.

At RAND Bales expanded his purview. Now, he claimed, he was not only capturing a miniature "society" itself within his experimental design but any and all communications systems. With human-engineering specialist Kennedy he compared the features of an air defense network—"a more or less typical large scale communication and control system"—to his own stripped-down interaction process.[50] Three air-defense problems— surveillance of the air picture, identifying friendly or unknown tracks, and

controlling fighters sent out—they found to be "on an abstract level" some-what similar to those problems Bales's task-oriented committees encoun-tered. Also, the two noted structural parallels: "It appeared that the step-wise operations involved in the total organization, as well as in component sections, could be tolerably well described as an interlocking series of some seven types of information-processing operations."

How hard, and at what an elevated level of abstraction, Bales had to work in order to find such commonalities! It is as if, through this process, the special room itself was becoming more and more abstract—no longer holding a tangible task but a standard one, no more a physical room so much as an all-purpose event space. Even the moniker "special" has been stripped away in favor of the generic stamp of "laboratory." Here was a proper environment in which a rational actor could act. Based on the air-defense comparison, a "symbol-aggregating-and-transforming process" be-comes the proper descriptor for the process whose endpoint results in what laymen call a "decision." It is "obviously a very complicated affair"—and its maintenance through acts and behaviors creates a common culture, a work-ing organization, and a viable communications system (figure 4.8). As Bales

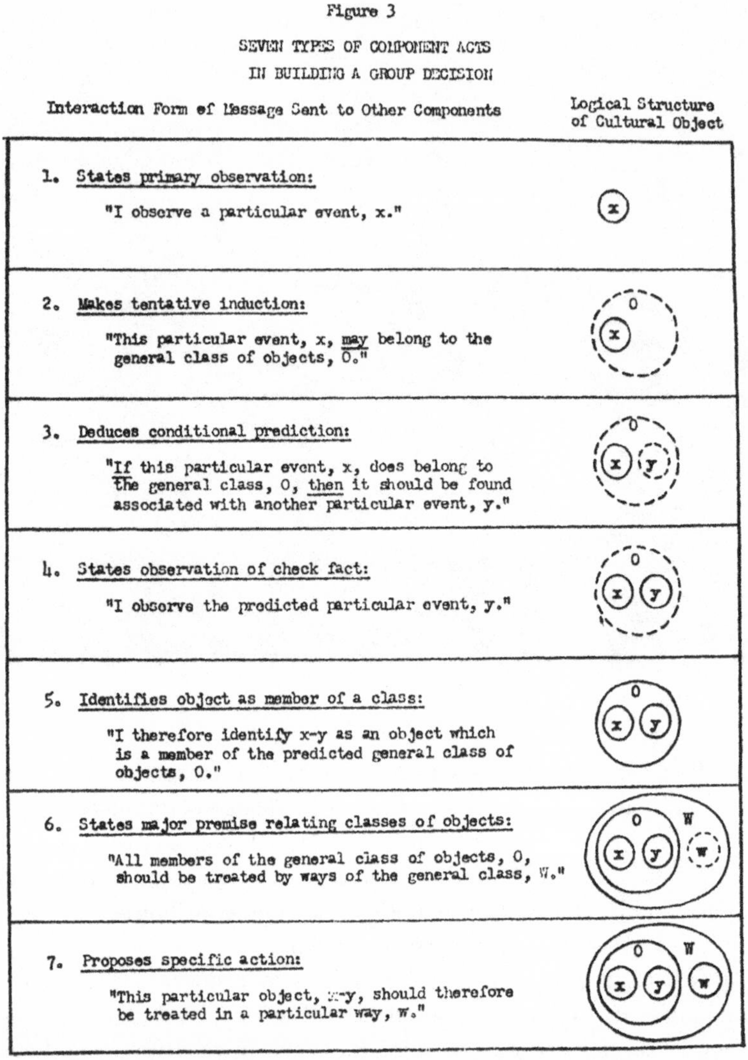

Figure 3

SEVEN TYPES OF COMPONENT ACTS
IN BUILDING A GROUP DECISION

Interaction Form of Message Sent to Other Components	Logical Structure of Cultural Object
1. States primary observation: "I observe a particular event, x."	
2. Makes tentative induction: "This particular event, x, may belong to the general class of objects, O."	
3. Deduces conditional prediction: "If this particular event, x, does belong to the general class, O, then it should be found associated with another particular event, y."	
4. States observation of check fact: "I observe the predicted particular event, y."	
5. Identifies object as member of a class: "I therefore identify x-y as an object which is a member of the predicted general class of objects, O."	
6. States major premise relating classes of objects: "All members of the general class of objects, O, should be treated by ways of the general class, W."	
7. Proposes specific action: "This particular object, x-y, should therefore be treated in a particular way, W."	

Figure 4.8. The seven steps identified in Bales's human interaction process analyses and in military defense networks. These steps were postulated to be universal to all communication and control systems (Bales 1954). (Reproduced with permission of RAND Corporation.)

emphasized, the general seven-step process was "not a formally perfect chain of logic, but rather . . . a set of symbol transformations which help to guide, though in an imperfect way, a process of decision-making behavior."

In sum, at RAND Bales built a tiny microcosmic system, one that was, for him, emblematic of all systems of communication and control. He came round to cybernetics in this way. "This fallible process," as he calls his isolate, is one in which a "slow feedback of accumulating error makes control possible." Control emerges out of error—and so Bales emphasizes fallibility, irregularities, and imperfections which spur the creation of stability. Such "failures" function like irrationality itself: for rationality (recall Simon's bowl holding the molasses) is best understood at the moments when it must contain irrationality. The successful end of a group interaction session is consensus: over the course of the meetings, the rates and ratios of positive, negative and information-seeking exchanges wax and wane according to predictable patterns, finally reaching equilibrium. A successful meeting, having passed the "bottleneck" of the decision point, ends in laughter and joking. An unsuccessful meeting does not, but it still follows an identifiable and even quantifiable pattern. Social relations (as a whole) and emotional states (of individuals) are restabilized when the seven steps work—either in missile defense or office headache-solving. No surprise that Bales collaborated too with mathematician Merrill Flood, also at RAND, to program a super-ego and the patterns of mental illness into the "Rational Mind" of a mechanical deliberator (as we glimpsed in chapter 1).[51]

4.5. Of Strange Twilights and Half Worlds

At the height of the Cold War, at a July 1958 press conference, President Dwight D. Eisenhower spoke of a murky "half world" in which Americans found themselves, neither fighting a full-on war nor not fighting a war. "We are not going to the ultimate horror," he said, stumbling a bit over his words, but neither were Americans experiencing "what we'd like to call a normal life."[52] An unusual state somewhere between war and peace, between the horrible and the humdrum, prevailed. In their own way, many American social scientists also captured this "between" state. They developed what may be called half-world sites in their research designs. In fact, the head of the Laboratory of Social Relations, Stouffer, used strikingly similar language some years earlier: he and his colleagues were experimenting, he said in his annual talk to the sociology meeting, in "a strange half light which is neither peace nor total war."[53] In that light, it seemed reasonable to pull everything possible into the inchoate struggle, in which military and socio-psychological purposes intertwined.

Today, when we pull back and survey a range of methodological experiments that took place in versions of what we are calling "the situation," one can find commonalities in other projects in the American behavioral sciences, projects that included elements of cultural anthropology, social psychology, psychoanalysis, ethology, child rearing studies, and operations research. This is not to claim that all of these disciplines, subfields, and theoretical culs-de-sac were the same at some basic level or were engaged in identical projects during the postwar years, but rather to suggest that there was to some degree a shared imperative, a shared set of assumptions about the *scientific process as it applied to human and social relations*, and, as a result, a shared product that consisted of a newly reworked form of stripped-down rationality—akin to Simon's bounded rationality and exercised in certain optimized physical spaces. Again, not all behavioral scientists shared this imperative; but those who did, did so with real enthusiasm. To adopt it was a choice and, very often, a lifelong commitment. These kinds of contributions might, on second thought, be called the "Integral Social Sciences," after a comment by Kenneth Boulding in *The Image*. Perhaps even better, they qualify as basic social science, whose practitioners built or prepared to build analytical models for the study of social systems.[54] On the road to a such a social science—social science as true science—were certain signposts. To carve out a situation and then render in code (eventually) everything that happened there was one of the most important on the way to the goal.

The push for a general theory accounting for every kind of human behavior at all times, as scientifically sane as this may have sounded, often sent research projects down odd byways: measuring gossip sessions; coding elements involved in planning a child's birthday party; categorizing types of Rorschach tests on one island while, nearby, another (albeit very small) island combusts into nothingness. With hindsight, one sees how strange these single-minded attempts to grasp human and social relationships in Lilliputian research situations could easily become. Method touched on madness. The strangeness was not so much in the situation as outside, where the designer overlooked his design, a dollhouse world in which the unknown and the irrational could purportedly be bound, understood, and possibly used.

Situation-based research was central to Cold War rationality.[55] Like the Cold War rationalizers, it was not based in any particular institution, although many of its most enthusiastic and specific articulators spent time at CASBS or RAND. It was a free-floating construct that was itself an index of the postwar rise of what intellectual historian Joel Isaac recently characterized as the "Tool Age" of the American social sciences.[56] Such methodological enthusiasm had its upside and its downside. The up: it made things possible. The down: it oversimplified. It reduced. It reified. It created fictions. It stipulated that a dollhouse world was a stand-in sufficient for this world, at least for operational purposes.

Out of such situations issued a peculiar and short-lived yet powerful new vision of social order. It was based on a stripped-down, judgment-free, methodologically sound form of rationality. No single person possessed such rationality, not even the overseer; rather, it circulated in the way the situation was set up. Bales located it in his special room when its proceedings were properly conducted by its scientific leader (as he later hastened to specify, a genial, democratic leader); Talcott Parsons and other eminent sociologists and psychologists embraced it. Call it situation-based rationality: it is, quite literally, bounded. Among other things, Bales's work might be seen as a real-life model of Simon's bounded rationality, described in chapter 2, for the actors within the special-room situation have only incomplete information about alternative courses of action. They are confined within the situation. Out of their shared constraints emerges a decision-making system and a guide to strategic action.

More pressingly, it showed that the assumptions of Cold War rationality, born in nuclear struggles and postwar geopolitics, could filter into and through the most intimate models for remaking the social world.

To return to our opening question of this chapter why did researchers go to such lengths to build their situations? To put it one way, borrowing from Harry Harlow, the ingenious deviser of the "strange situation" in primate research, it was "to see . . . with special clarity."[57] All sorts of situations, or perhaps *the situation* itself, had the potential to become mobile, detachable on-the-ground test sites where all kinds of phenomena could be better examined and new methods could flourish. Special powers of seeing into and penetrating the otherwise mysterious phenomena of human behavior resulted: this was the advantage of building a microcosmic world.

In the next chapter we will see more of exactly what the bowl was like: it could take the form of a war room, for example, a conference room, or more specifically the White House "situation room." Or, it could be a nuclear standoff imagined via simulations. It could contain a Strangelovian crisis and its algorithmic solution. Or it could hold a game—of a different sort.

FIVE

World in a Matrix

By the fall of 1969, the Americans and North Vietnamese had reached stalemate in their negotiations to end the Vietnam War. Richard Nixon had entered office the previous January promising to end the conflict, yet American GIs continued to return home in body bags. Meanwhile, at negotiations in Paris, the North Vietnamese steadfastly refused concessions to the Americans. Nixon exuded fury—and wanted to be sure the Vietnamese and their Soviet allies knew it. "I call it the Madman Theory, Bob," he told his chief of staff, H. R. Haldeman, who would later serve time in the wake of the Watergate scandal. "I want the North Vietnamese to believe that I've reached the point that I might do anything to stop the war"—anything, including escalating to an all-out nuclear exchange.[1] Thus, toward the end of October, the air force launched the first waves of a massive airborne exercise, "Giant Lance," in which nuclear-armed B-52 bombers prowled the skies over the polar ice cap, toward and away from the Soviet Union. While their presence was designed to be apparent to Soviet military observers—who certainly would alert the Soviet leadership to the American threat—the flights were carefully hidden from the view of both the American public and allies around the world.[2] Was this truly madness, or did it reflect a more calculated kind of lunacy? For nuclear strategists and political scientists of the 1960s, steeped in the rational-choice vernacular employed by the likes of Herman Kahn (chapter 3), it must have seemed like evidence of the latter. In particular, Nixon's bizarre logic of international relations has often been laid at the doorstep of game theory, a mathematical theory of interaction between "rational" individuals (defined in a sense peculiar to the theory) that had initially been developed by John von Neumann and Oskar Morgenstern in their 1944 *Theory of Games and Economic Behavior*.

Kahn had made qualitative reference to the game of "Chicken" in his analysis of escalation and deterrence, suggesting that calculated, rational choices could be made at each step of the escalation ladder, and during the 1960s formal elements of game theory became increasingly intertwined with discussions of war and peace. In his influential book, *The Strategy of Conflict* (1960), economist Thomas Schelling drew on game theory to analyze "the threat that leaves something to chance," the introduction of an element of randomness into the bargaining mix of carrots and sticks employed by a negotiator, since actually following through on a threat, or transparently failing to follow through, carries a strategic cost. The use of randomness in such situations (the employment of so-called mixed strategies by game players) was an essential ingredient of game theory from some of its earliest formulations by Von Neumann in the 1920s. As the "rational-actor perspective" employed by Schelling spread across academic political science in the following years, the Cold War was reinterpreted as a specific kind of game: a dilemma game in which rational calculation failed to achieve rational outcomes, and conversely, in which apparent "irrationality" became a rational strategy. Subsequent histories of Cold War nuclear policy and intellectual culture, both popular and academic, have reiterated the connection between game theory and this kind of thinking.[3]

Yet the emergence and persistence of game theory at the center of debates over nuclear strategy, arms control, and international diplomacy presents a puzzle in light of the foregoing chapters. We have seen how a rational choice approach to the study of nuclear strategy quickly proved empirically inadequate during the 1950s and 1960s and how a richer understanding of psychology, grounded in theories of "cognitive dissonance" was called for to supplement rational calculation as the basis for decision making. Yet again and again, Cold War intellectuals turned to the spare notations and logic of game theory to tackle problems of strategy and arms control. In 1967, even as the limitations of rational calculation as a tool for addressing such problems were becoming clear in the work of Charles Osgood and others, researchers at the University of Pennsylvania's Management Science Center (along with groups at a number of other universities) could present game theoretic models of the Vietnam War to their patrons at the Arms Control and Disarmament agency that would capture precisely the dilemma facing Nixon: whether to escalate the war in the quest for victory while possibly provoking Vietnamese escalation and an even bloodier stalemate (figure 5.1).[4] That the Cold War was literally a game in this stripped-down, spare sense remained a consistent point of departure for discussions of arms races and

	V.C.	
	NOT ESCALATE	ESCALATE
U.S. — NOT ES-CALATE	3 3 De-escalation	1 4 V.C. Victory
ES-CALATE	4 1 U.S. Victory	2 2 Escalation

Figure 5.1. Nixon's dilemma, as imagined by Arms Control and Disarmament Agency contractors, 1967. (Russell L. Ackoff et al., *A Model Study of the Escalation and De-Escalation of Conflict*, Report to the US Arms Control and Disarmament Agency under Contract ST-94, Management Science Center, University of Pennsylvania, March 1, 1967, 54.)

nuclear war, even as the adequacy of game theory's calculating brand of rationality for dealing with such situations came in for criticism.

To understand the persistence of such games in Cold War strategic debates (not to mention far-flung corners of the sciences) this chapter explores several episodes in the history of the particular game captured in the matrix above, better known as the prisoner's dilemma. According to one influential taxonomy, the prisoner's dilemma is only one of seventy-eight distinct two-person non–zero-sum games; the game of "Chicken" referenced by Russell and Kahn is another.[5] Yet from its initial formulation by mathematicians working on behalf of the United States Air Force in the early 1950s, no other game has been so commonly associated with the paradoxes of security in the age of nuclear weapons. Specifically, for many individuals, the prisoner's dilemma focused attention precisely on the divergence between the maximizing rationality embodied in mathematical programming models (see chapter 2) and the more complex kind of rationality needed to achieve substantively rational national security outcomes in an arms race. And while a number of new mathematical results would emerge during the later 1970s and 1980s that would clarify the circumstances under which rational calculation might produce substantively rational outcomes in the context of prisoner's dilemma games, at least during much of the period treated here, it was not clear that these results constituted a definitive "solution" to the problems of the arms race.[6] Despite the high hopes of its early practitioners (and the theory's ability to rationalize a certain amount of calculated randomness), game theory, per se, did not immediately provide anything like a straightforward calculus for solving "the problem of the bomb."

Instead, this chapter focuses on another aspect of game theory that may help explain the theory's enduring attraction even to individuals who deemed that rational calculation was insufficient for—or even irrelevant to—the study of arms races, conflict, and cooperation: the way that its notational devices and conceptual framework provided an exceptionally flexible and adaptable set of tools for coding and thinking about behavior in a wide range of disciplinary contexts. The game matrix in particular served a key function in the research strategy of social and behavioral scientists in this period. As we saw in the previous chapter, postwar behavioral scientists found it profitable to focus on particular microcosms in which to study human behavior—different "situations," whether that meant a laboratory, a room glimpsed through a one-way mirror, an island, or a particular social encounter—but getting from observed behavior in the microcosm to supposedly general insights into the nature of human social interaction and choice behavior requires a certain narrowing of vision, a stripping away of aspects of a situation deemed nonessential, and the preservation of those that appear essential. Stripping a situation down to payoffs in a game matrix allowed game-theoretic rationality to vault between contexts and across disciplinary lines and spatial and temporal scales. In the process, Nixon's Vietnam War became a prisoner's dilemma game, played out between the superpowers—but so too did the interactions of human laboratory subjects, economic actors, and even insects undergoing natural selection. With these situations encompassed by a common matrix upon which rational calculation might operate, the game-theoretic strain of Cold War rationality could become a point of reference for debates unfolding in a surprising diversity of contexts.

5.1. Game Theory and Its Discontents at RAND

The earliest version of the game now known as the prisoner's dilemma was devised by mathematicians Merrill Flood (whose experiments on bargaining we encountered in chapter 1) and Melvin Dresher at the RAND Corporation in January 1950. At that time they performed a brief experiment in which two players repeatedly played a game that they originally called "a non-cooperative pair," with pennies as the payoffs.[7] By the spring of 1950, the game had acquired the story with which it is now commonly associated, given to it by the Princeton mathematician Albert Tucker, who was trying to explain the game to an audience of psychologists at Stanford University.[8] One of the earliest versions of this story found in Flood's papers—

titled simply "A Two-Person Dilemma"—bears Tucker's name and dates to May of that year. It runs as follows:

> Two men, charged with a joint violation of law, are held separately by the police. Each is told that
> (1) If one confesses and the other does not, the former will be given a reward of one unit and the latter will be fined two units.
> (2) If both confess, each will be fined one unit.
>
> At the same time each has good reason to believe that
>
> (3) If neither confesses, both will go clear.

Or, captured concisely in game theory's "payoff matrix" (figure 5.2):

	confess	not confess
confess	$(-1, -1)$	$(1, -2)$
not confess	$(-2, 1)$	$(0, 0)$

Figure 5.2. "Payoff matrix" for the prisoner's dilemma.

Here, the matrix entries (e.g., $(-2, 1)$) denote the payoffs, in some arbitrary unit of value, to the row player and the column player respectively. Thus the "prisoner's dilemma" came into existence.[9]

The "dilemma" in question is often taken to refer to the decision facing the two prisoners: should I confess, locking in a one-unit fine, or should I stay silent in hope of gaining freedom, only to risk a fine of two units if the other prisoner turns state's evidence? And given that Flood, Dresher, and Tucker were writing against the backdrop of some of the darkest days of the Cold War, it is also hard not to see reflected in their story a dilemma facing capitalists and communists, East and West, and the logic of the arms race and military escalation that would ensue in the absence of trust between the players, or their inability to produce enforceable agreements.[10] Yet the Cold War did not immediately become a prisoner's dilemma: instead, the dilemma that concerned the RAND mathematicians had less to do with the one facing the prisoners (or the superpowers), and more to do with the challenge this particular game posed for their attempts to build a theory of games that would suit the needs of their military patrons. Their forays into experimental behavioral science notwithstanding, Flood and Dresher were first and foremost mathematicians, formulators of axioms and provers of theorems. Moreover, the direction of their mathematical interests was intimately connected with the status of game theory as a branch of applied mathematics in the late 1940s, and the nature of the intellectual agreement reached between practitioners of game theory and the postwar air force.

Although it only gained wide circulation with the publication of Von Neumann and Morgenstern's *Theory of Games and Economic Behavior* toward the end of World War II, game theory was initially developed largely outside of the military context. In their book, Von Neumann and Morgenstern sought to establish games (like poker or chess) as the fundamental unit of analysis for a new social science that would address a number of shortcomings of traditional economic theory. Applying logic to mathematical axioms of "rational behavior" in game situations, Von Neumann and Morgenstern's theory sought to "solve" games. In their view, a "solution"—"a characterization of 'rational behavior'"—would ideally consist of a "complete set of rules of behavior in all conceivable situations."[11] Yet despite the hefty size of *Theory of Games* when it appeared in 1944, the only part of the theory that came close to realizing this vision dealt with two-person zero-sum games, that is, games in which the winnings of one player were the losses of the other. In such situations, the principle of "rationality" to be applied was fairly straightforward: choose a strategy that will maximize your expected winnings while simultaneously minimizing your opponent's expected winnings (a so-called minimax strategy). The key to "solving" such games was Von Neumann's insight that a player's rational strategy might not be to follow a determinate course of action, but to choose an action at random according to a probability distribution. If these kinds of randomized strategies ("mixed strategies") were available to players, Von Neumann could prove that "rational" strategies existed. If rationality meant maximization, the calculated use of randomness made rationality possible.

Yet even in the relatively simple situation of the two-person zero-sum game, Von Neumann only proved that solutions *existed*, rather than providing algorithms for the actual calculation of courses of action. The theory of games involving more than two players, of situations where bargaining over surpluses is possible, remained still more fragmentary. For these games, Von Neumann and Morgenstern had suggested that the players would form coalitions to win and somehow divide the spoils of their collusion, subject to the constraint that individual players might "defect" to demand a greater share of the winnings elsewhere. This part of the theory provided nothing like the "complete set of rules of behavior" that Von Neumann and Morgenstern had hoped for from the outset: solutions consisted of *sets* of possible payoff distributions to players; they provided little guidance to players on how to proceed; and Von Neumann could not prove that all games possessed such solutions.[12]

This situation was problematic because game theory's appeal to the military was built, at least in part, on the promise of solving games to find determinate rules of strategic interaction. As we saw in chapter 2, this was also the promise held out by the mathematics of linear programming and numerical methods of optimization, which also attracted significant military funding and interest during this period. Game theory and linear programming are in fact closely related, as George Dantzig discovered when he visited Von Neumann at Princeton in 1947. During their meeting, Von Neumann conjectured that the problem of solving two-person zero-sum games and the linear programming problem were equivalent: in this instance, the game player, like the air force comptroller's office, sought to maximize a linear function subject to a linear system of inequalities. This equivalence was further explored by Albert Tucker and his students at Princeton in subsequent years.[13] The mathematics of the two-person zero-sum game thus became a key focus of attention for the community of mathematicians at Princeton and at RAND for a couple of interrelated reasons: not only was the problem of solving such games equivalent to practical problems of programming and logistics, but Von Neumann had already developed a fairly coherent understanding of what it meant to "solve" such games for rules of rational behavior. Therefore the bulk of game theory studies pursued at RAND and elsewhere focused on solving particular two-person zero-sum games, such as models of duels between fighter and bomber aircraft, or games in which commanders had to allocate scarce resources across multiple battlefields on the assumption that his opponent would make similar calculations.[14]

The connection between game theory and the military was further reinforced in the later 1940s by the development of computers and algorithms for actually finding practical solutions to two-person zero-sum games and linear programs. While we have seen that truly "optimal" solutions to such problems were not necessarily attainable given computing's state of the art, the computer's calculational abilities remained a constant point of reference for those interested in solving games and related optimization problems. By 1950, this point of reference was close at hand indeed: RAND had acquired a commercially available analog computer in 1948, and a year later, corporation mathematicians began scouting the possibilities for constructing their own electronic digital computer, which would become operational in 1953.[15] Computers could even potentially be called on to mimic the human tactic of "bluffing" (or Nixon's unpredictable "madman" tactics) through the creation of algorithms for generating game theory's "mixed strategies," strategies where a game player chooses a course of action at random. Already in the spring of 1947, the RAND Corporation had developed a device that would convert inputs from a physical "random frequency pulse source" into randomly distributed digits printed on IBM punch cards. Within a few years, one of the mathematicians involved in the project could look forward to the day when improved numerical processes and improved computational power "will permit us to compute our random numbers as we need them." The computer—whether as concept or as material reality—had the potential to serve as game theory's ideal rational agent, bringing both mechanistic calculation and Nixon's seemingly erratic bluffs within a common ambit.[16]

But despite these rapid successes, there was a growing recognition of the problems that lay beyond the two-person zero-sum game. As the RAND mathematicians noted in research memorandum after research memorandum throughout the late 1940s, Von Neumann and Morgenstern's method for analyzing non–zero-sum games needed reassessment. One problem identified quite early concerned the formation of coalitions. As one mathematician put it in a 1948 report on the state of game theory at RAND, "the assumption may be considered utopian" that players would form coalitions in many instances, and he called for the investigation of games in which there was no possibility of coalition formation.[17] But more troublesome still was the fact that Von Neumann and Morgenstern's solution seemed incapable of prescribing "rational behavior" in the same way that the theory of the two-person zero-sum game had done so clearly. Their solutions, as Albert Tucker and Duncan Luce wrote in 1959, "seem neither to prescribe rational behavior nor to predict behavior with sufficient precision to be of empirical value." The problem of practical reasoning—how to decide what one should do in any given situation—could not simply be reduced to rational calculation.[18]

As a result of these shortcomings, a number of alternative "solution concepts"—alternate paths to the holy grail of "solving" a game—emerged among the RAND-affiliated game theorists in the late 1940s and early 1950s. Perhaps the most sweeping attempt in this regard stemmed from the work of John Nash, then a Princeton graduate student who spent summers at RAND in the late 1940s and early 1950s. Nash's vision for game theory distinguished between "cooperative" games (the focus of most of Von Neumann and Morgenstern's work) and "noncooperative" games, in which players act "without collaboration or communication of any sort."[19] Instead of modeling the formation of coalitions (which would eventually break up anyway as each player clamored for his share of the gains from the collaboration) Nash assumed from the outset that individuals would apply to non–zero-sum games the same principle of rationality-as-optimization that had worked so well in the context of two-person zero-sum games (where communication between the players was pointless). By this logic, players would seek a strategy that "maximizes [the player's] pay-off" on the assumption that their opponents were simultaneously doing the same thing. The resulting sets of strategies—there could be more than one—were "equilibrium points."[20]

Nash's solution concept was a key piece of the context in which Flood and Dresher performed and analyzed their first game experiments at RAND. Here is how they would have viewed the "two-person dilemma" in the spring of 1950:

> Clearly, for each man the pure strategy "to confess" dominates the pure strategy "not to confess." Hence, there is a unique equilibrium point given by the two pure strategies "to confess." In contrast with this non-cooperative solution one sees that both men would profit if they could form a coalition binding each other "not to confess."[21]

The term "equilibrium point" refers to Nash's noncooperative solution concept, which would seem to suggest a strategy of mutual confession—thereby locking in a suboptimal outcome for both players. This stands at odds with the kind of solution Von Neumann and Morgenstern might have proposed, that is, "form a *coalition* binding each other 'not to confess.'" (Or, as we saw in chapter 1, to babysit.) Thus, the principal "dilemma" in question was not the one facing the prisoners, but the one facing the RAND mathematicians seeking to develop a comprehensive theory of multiplayer and non–zero-sum games. In effect, the prisoner's dilemma began its existence as something like a glorified mathematical counterexample.

It is not clear how successful the RAND experiments were in resolving the dilemma of which solution concept to choose for solving non–zero-sum games. Flood concluded that the experimental subjects showed "no tendency to seek as the final solution . . . the Nash equilibrium point," but neither did they cooperate in a straightforward manner.[22] The result would certainly have pleased Von Neumann, who had never felt the Nash equilibrium concept particularly interesting or appealing.[23] Nash, for his part, felt that the experiment did not constitute an adequate test of his equilibrium concept. His objections—recorded in a 1952 research memorandum written by Flood—hint at fundamental problems facing any attempt to experimentally verify any theory of games. To generate statistically meaningful data, the experimenters needed to repeat the game multiple times; however, since players have memories, subsequent games are effectively not the same game as earlier ones. One possible solution to this problem would be to have players rotate in and out of the game at random so that they could not get to know one another.[24]

But Nash's proposal begs the question: what was the point of a theory of games in the context of RAND and the needs of the air force? Was it intended to capture some essential feature of how people really do play games in some idealized and perfectly controlled situation that was probably impossible to create in a RAND Corporation office, much less on the battlefield? If not, then the point of further experimentation on the prisoner's dilemma would seem unclear. And indeed, experiments on games appear to have died out by the mid-1950s in tandem with a decline in enthusiasm for game theory more generally at the corporation. The decline doubtless had many causes, from budget cuts to the impact of the McCarthy security hearings on the RAND staff. However, one cannot help but imagine that the intellectual and methodological differences of opinion exposed by the project of "solving" non–zero-sum games might have played a role as well. Game theory in the hands of its military patrons was intended as a guide to what *should* be: part of a program to improve (if not optimize) the use of weapons systems or the functioning of supply chains. Decision making, guided by the solution of linear programs or tactical games via computer program or servomechanism, was intended to bypass the "human factor" as much as possible, rather than embrace it as an essential ingredient of rationality itself.

5.2. Mathematics' Loss Is Psychology's Gain

At precisely the time that the RAND mathematicians seemed stymied in their quest to "solve" the prisoner's dilemma, the game was beginning to gain an ardent following in social psychology. This fact might seem odd when viewed in light of traditional models of the relationship between "theory" and "application," yet it makes a great deal of sense given game theory's peculiar place at the intersection of mathematics and the study of human decision making. Like his intellectual forebears, the mathematical logicians of the late nineteenth century, who had striven to liberate logic from the clutches of the psychologists, Von Neumann drew attention to what he saw as the exclusion of "psychology" from his analysis of the two-person zero-sum game, since the players' courses of action could truly be determined by calculation alone.[25] By the same token, for at least some mid-century psychologists, it was precisely the failure of rational calculation to solve the prisoner's dilemma that made the game so interesting. Instead, the prisoner's dilemma (and its signature game matrix) offered a structured, controlled template for producing psychological knowledge, much as did the simplified "situations" found in the work of R. F. Bales and other social psychologists during this period. The pursuit of such knowledge—far more than the "theory" of the RAND mathematicians—would ultimately help insert the prisoner's dilemma into discussions of the dilemmas of international arms control in the thermonuclear age (figure 5.3).

While many studies of games and game-playing behavior would appear in the 1950s and 1960s, probably the most exhaustive experimental exploration of the prisoner's dilemma is recounted in Anatol Rapoport's *Prisoner's Dilemma: A Study in Conflict and Cooperation* (1965). Rapoport was in many ways the ideal person to bridge the gap between game theory as practiced at RAND and psychology during this period: he held a PhD in mathematics and also had spent time at the University of Chicago's Committee on Mathematical Biology in the late 1940s and early 1950s, eventually moving to the University of Michigan in 1955. Rapoport first encountered the prisoner's dilemma during a year's sabbatical in 1954–1955 at the Center for Advanced Study in the Behavioral Sciences at Stanford, in a seminar led by the mathematician and measurement theorist R. Duncan Luce. According to his autobiography, Rapoport quickly saw the implications of the game for thinking about patterns of conflict and cooperation, both among individuals and nations. Upon his arrival at Michigan the next year, he embarked on a multiyear study of human teamwork and cooperation for the air force, which was interested in improving the performance of its flight crews. In the course of these studies, Rapoport began a series of experiments to measure individuals' tendencies to cooperate in the prisoner's dilemma game and how this tendency changed over time—not to test any particular theory of games, but because "cooperation" in this context seemed a good proxy for "teamwork," Rapoport's (and the air force's) true variable of interest.[26]

Prisoner's Dilemma thus offers a perspective on the significance of the prisoner's dilemma very different from that found in Flood, Dresher, and Tucker's work. The game was significant not because it possessed a generally accepted "solution" grounded in rational calculation, as the mathematicians had hoped to find, but precisely because it did not; or as Rapoport would put it, "the potentially rich contributions of game theory to psychology will derive from the failure of game theory rather than from its

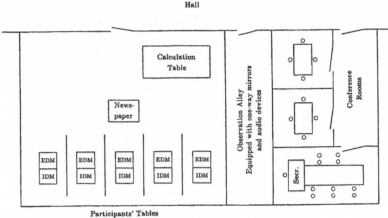

Figure 5.3. Putting human subjects in a matrix in the course of a
geopolitical simulation game. (Harold Guetzkow, "A Use of Simulation
in the Study of Inter-National Relations," *Behavioral Science* 4 [1959]: 189-91.)

successes."[27] Rather than focusing on calculation, one very narrow aspect
of human reasoning, observations of prisoner's dilemma–playing behavior
would help develop insights into what Rapoport called "real psychology":
"the realm of personality, intellect, and moral commitment," which could
be explored quantitatively by looking at the frequency with which individuals cooperated in response to varying payoffs and other experimental
conditions.[28] Not surprisingly, while the theory of the two-person zero-sum
game had been central to the RAND mathematicians, Rapoport found such
games of marginal interest: with their strategies determined by rational calculation, they could only offer psychological insight inasmuch as players
might "irrationally" depart from the theoretical predictions.[29]

Despite Rapoport's hopes for the role of prisoner's dilemma in psychology, his results actually seemed to say relatively little about "real psychology." Partly, this may have reflected the demands of experimental rigor. In
order to eliminate hard-to-control verbal or gestural interactions, communication between the players was banned, so that many of the subtle path-
ways by which the subjects might have communicated intentions or values
were closed off. Like Flood and Dresher before him, with a few exceptions
Rapoport also focused principally on analyzing multiplay runs of prisoner's
dilemma by the same players in order to gain statistically significant data

(rather than, for example, comparing the behavior exhibited by members of different populations of players in a single play of the game). For example, he analyzed the differences between the ways males and females played the game, yet concluded in general that "whatever individual differences exist among the players (and it is difficult to believe other than that they exist) tend to be ironed out in the course of the interactions between them," so that much of the variation in outcomes "is accounted for not by the inherent propensities of the players to cooperate or not cooperate, but rather by the characteristic instabilities of the dynamic process which governs the interactions in Prisoner's Dilemma." The effects of repeated interactions and the payoffs at stake thus seemed more significant than preexisting characteristics of the individual players, such as intelligence or personality, in determining outcomes.[30] While there was a slightly greater tendency overall toward cooperation than noncooperation (with the overall frequency depending mostly on the structure of the payoffs involved), the most impressive result of the trials was an overwhelming tendency for players to learn "conformity," even if this was not necessarily the same thing as "teamwork." Players either cooperated most of the time or did not cooperate, so that "typically, toward the end of the sessions over ninety percent of the responses are matched."[31]

If Rapoport focused much of his attention on interaction processes rather than personal qualities, the opposite was true of another pioneer of prisoner's dilemma laboratory studies, Morton Deutsch. Deutsch's much-cited experimental study of the prisoner's dilemma, "Trust and Suspicion," appeared in the *Journal of Conflict Resolution* in 1958, making it one of the first experimental prisoner's dilemma game studies to appear since the RAND studies from earlier in the decade. Like Rapoport, Deutsch represented a very different disciplinary lineage than Flood, Dresher, and Tucker, completing his PhD thesis in experimental social psychology at MIT's Research Center for Group Dynamics in 1948.[32] From its roots in the work of Kurt Lewin and his students in the 1930s, group dynamics had focused on developing an experimental understanding of the interaction between individual personality and social environment, most notably the relationship between personalities and leadership styles on the one hand, and group productivity in the workplace and in civic life on the other. Lewin's classic study during this period presented observations of social interaction in two groups of fifth- and sixth-grade children who were brought together around craft activities. Comparing observations on the behavior of the groups under different styles of adult leadership—one "democratic," participatory, and consultative and the other "authoritarian," characterized by top-down leadership—Lewin clearly thought he had found convincing evidence for the superiority of democratic leadership. The authoritarian group exhibited greater social tension, hostility, and scapegoating behaviors; the democratic group was characterized by greater intragroup communication, stability, and productivity.[33]

During the 1940s and 1950s this tradition of research attracted substantial support from both the military and industry, which valued insights into teamwork, and from reformers interested in resolving social conflicts. Correspondingly, the goal of research in group dynamics was the use of motivational training techniques to induce social and behavioral change, whether on the factory floor, in combat teams, or in housing developments.[34] During the 1950s, Deutsch in many ways epitomized this intellectual tradition. In addition to holding a faculty position, during 1952–1954 he was a member of the Committee on Civil Rights of the Society for the Psychological Study of Social Issues, in connection with his work studying interracial housing in New York and Newark, New Jersey.[35] Much like Rapoport with the air force, he also received funding from the Office of Naval Research for his experimental research into conditions promoting cooperation in small groups.[36]

Deutsch's landmark 1958 study of trust and suspicion clearly drew on his work for the Office of Naval Research and focused on understanding the conditions that would foster trusting attitudes in a small group setting. "Trust" in this instance was not simply a matter of cognition, of successful prediction of future events; it also involved the positive and negative "motivational consequences" of confirmation or disconfirmation of belief. Deutsch hypothesized several factors that might increase the "individual's confidence that his trust will be fulfilled" relating to the perception the individual had of others. These included "the nature of the intentions that the individual perceives his potential object of trust to have; the perceived power of the object of trust to cause the desired events; the power relationship between the individual and his object of trust; the influence of communication upon the development of trust; the influence of third parties upon the development of trust between two people; the individual's self-esteem as it affects his readiness to trust." Experimental plays of the prisoner's dilemma game, he argued, would provide the perfect opportunity to test the conditions that might reinforce trusting behavior since "The essential psychological feature of the game is that there is no possibility for 'rational' individual behavior in it unless the conditions for mutual trust exist."[37]

Here, Deutsch quite naturally equated "cooperation" and trust with rational behavior, with outcomes that are best for the "team" of players as a whole, and "motivational consequences" with the psychological impact on individuals of wins and losses. With these equivalences in place, Deutsch thus proceeded to test several possible factors that might create trust. For example, he performed experimental trials of the game under three different "motivational orientations": "cooperative," "individualistic," and "competitive." Each orientation was imparted before play via verbal instructions to the subjects "which characterized . . . the objectives they were to have in playing the game and the objectives they could assume their co-player would have."[38] Perhaps not surprisingly, the "cooperative" orientation instructions produced consistently high percentages of cooperative strategy choice, whereas a "competitive" orientation was nearly always lowest. Other experiments and observations focused on behavior in situations where communication was permitted or not permitted, with Deutsch observing that even players who were given the opportunity to communicate often did not do so effectively, in Deutsch's opinion. Deutsch thus concluded by proposing future research that would examine how different opportunities and avenues for communication stimulate trust between the players.

Deutsch's analysis thus is striking in the richness of social interaction and social roles that he sought to investigate, drawing in considerations of cooperation, communication, power, and social connectedness. His sense of rationality, far more expansive than the rationality-as-optimization pursued by the RAND mathematicians or indeed the conformity discovered by Rapoport, was nevertheless in some ways oriented toward similar ends. Even if the goal was not to axiomatize reason, to reduce it to a set of rules and calculations, experts were still needed to engineer the motivational environment in which groups of individuals could come to behave rationally (in this case, to cooperate). Rationality would be generated not by computer but by some kind of collective therapy. But in the process of adapting games to the laboratory and to practical problems of mediating social conflicts, Deutsch's work became almost completely divorced from the "theory" of games in any sense that Flood and Dresher might have recognized.

Rapoport and Deutsch both began their psychological investigations of the prisoner's dilemma with funding from the US military, which was interested in understanding phenomena of teamwork and cooperation in small groups like the crews of airplanes or submarines. Quite quickly, however, both men came to imagine that the insights generated by their research had relevance to the arms race shaping up between the United States and the Soviet Union in the 1950s. The impulse to do this stemmed in part from their political backgrounds. Deutsch had been involved with antiwar causes since learning of the nuclear bombings at Hiroshima and Nagasaki; and as a result, his doctoral thesis on learning in cooperative and competitive environments (which would set the scene for much of his work on trust and suspicion) had its roots in the postwar years when he "had been more interested in world peace than in education," the ostensible subject of his thesis.[39] Rapoport's revelation was due in no small part to his personal convictions: a socialist, he had been a vocal opponent of the increasingly violent exchanges of rhetoric between the United States and the Soviet Union since the late 1940s. As a result, during 1954–1955, he was also part of a reading group that that met to discuss the work of Lewis F. Richardson, a Quaker meteorologist who had brought to bear statistical and mathematical models to study the progression of arms races and the outbreak of wars. Among other things, Richardson had written down differential equations describing the dynamic interactions between nations undergoing arms buildups. Depending on the parameters of the equation, increased arms expenditures in one country could lead to increased arms expenditures in the other, with overall armaments crescendoing in a chain of reactions and counterreactions.[40] Richardson's models took a page from the equations of classical physics or epidemiology, so that war, driven by moods for which "there are no rational components" is like a disease with a "regular, almost predestined course."[41] Nevertheless, according to Rapoport, "The connection between [the prisoner's dilemma] and the situation produced by the arms race occurred to me at once"; "cooperation" meant undertaking arms control, while noncooperation meant continued weapons development.[42]

Through such connections, the prisoner's dilemma could emerge as a promising tool for investigating problems of conflict and cooperation at the international as well as interpersonal levels, a far broader mandate than Flood and Dresher's at RAND. This connection would be reinforced by the development of new institutions and sources of funding in the later 1950s that sought to apply the results of behavioral science to understanding Cold War problems of peace and violent conflict. These included the University of Michigan's Center for Research on Conflict Resolution (and the in-house *Journal of Conflict Resolution*, in which both Rapoport and Deutsch published) and after 1961, the Arms Control and Disarmament Agency, which funded several projects involving game theory throughout the 1960s. In this context, the prisoner's dilemma *did* become a key theoretical framework for thinking about "the problem of the bomb," and the game matrix could leap from the mathematics of optimization to psychological laboratories to problems of war and peace writ large.

However, the lessons of the prisoner's dilemma would ultimately prove ambiguous for both Deutsch and Rapoport, just as they did for many other academics associated with peace research and conflict resolution. Game theory's spare description of reality allowed them to move seamlessly between human subjects in a laboratory and the affairs of nations—or so they thought. But their embrace of a single game matrix to stand in for a common set of problems also papered over intellectual fault lines that would reemerge periodically amid the turbulent politics of the later 1960s. Mathematicians and operations researchers working for the military or for the Arms Control and Disarmament Agency continued to search for "solutions" to games like prisoner's dilemma grounded in calculations of individual rationality, grasping for hope that arms races could in fact be averted in a world where nations behave like rational egoists and binding commitments are implausible. Meanwhile, despite embracing the basic notations of game theory to talk about arms races like their more mathematically inclined counterparts, the psychologists' laboratory studies of teamwork and cooperation seemed to lead in different directions altogether. Deutsch's work sought to empower the group psychotherapist or counselor to act upon attitudes of the parties to a conflict and somehow guide them toward rational (i.e., cooperative) behavior. Rapoport, by contrast, would prove less enthusiastic about such attempts to "engineer" rationality, preferring to focus on using prisoner's dilemma to demonstrate the possibility for (but not the necessity of) an enlightened, more empathetic logic of decision making.[43] All the while, the search for more secure solutions to prisoner's dilemma and to the problem of the arms race continued, spilling over into new disciplinary terrain.

5.3. Rational Outcomes without Intelligent Actors

Even if the social psychologists and their mathematician colleagues might have held different views on the efficacy of rational calculation for solving prisoner's dilemma games, they nevertheless seemed to agree that the game matrix was the grist upon which some kind of reasoning process would operate. It is therefore remarkable that by the 1970s prisoner's dilemma would come to be embraced by evolutionary biologists who were deeply skeptical that the game "players" they studied (whether animals or insects) were capable of anything like human cognition. Here again, game theory's notational systems provided an abstraction that proved capable of encompassing the clash of the superpowers in Vietnam, fighting deer, or egg-laying parasites—although as we shall see, fitting the natural world into a game matrix required somewhat more ingenuity than was needed for prisoner's dilemma to jump between prisoners and flight crews, or even between laboratory game-experiments and international politics.

The first intellectual shift needed to transform nonhuman organisms into realistic prisoner's dilemma players was to endow them with genetically determined self-interest. This was not an obvious thing to do given the dominant state of scientific opinion in the 1950s. Among biologists interested in animal behavior, a long intellectual tradition emphasized the natural origins of social cooperation rather than competition. Darwin's "struggle for existence" did not operate only or even primarily on the level of individuals; altruistic behavior, wherein individuals sacrificed themselves for the good of others, also evolved through its benefit to the species as a whole. The most fundamental drives in nature were toward social harmony and cooperation.[44] Aggression and competition certainly did occur in nature—for example, in fights for territory and mates—but they were usually carefully restrained to prevent needless slaughter. Indeed, ethologists would argue that in many instances, animal fights had become "ritualized," involving symbolic displays of threats and rarely resulting in serious violence.[45] Moreover, responding to postwar revelations of the horrors of Nazi racial science and eugenics, a number of prominent biologists came to question the connection between genetics and social behavior in humans, embracing culture as an explanation of behavior and reversing the earlier emphasis of eugenics that had sought to address social problems through monitoring and manipulation of human heredity.[46]

Several developments changed this situation in the 1960s. In the realm of popular culture, a slate of works by authors such as Robert Ardrey and Desmond Morris popularized new theories of the origins of human social behavior that restored biology over culture and put violence back into human nature. Ardrey's *African Genesis* (1961), for example, argued that in fact a growing lust for hunting, bloodshed, and weaponry had characterized the evolutionary transition from apes to humans.[47] In addition, a new generation of evolutionary theorists—most notably W. D. Hamilton, John Maynard Smith, George Price, Robert Trivers, and Richard Dawkins—emerged who were committed to restoring what they saw as Darwin's original emphasis on inheritance coupled with individual advantage as the engine of evolutionary change. Their theories embraced a new vision of life fit for the age of DNA: organisms are information-processing machines, programmed by instructions coded in their genes from birth.[48] These intellectual shifts raised challenges both for those who would privilege social interactions over biology as a force shaping behavior, and for those who explained adaptations by reference to innate social tendencies or to their contribution to the "survival of the species." The problem of how to reconcile the neo-Darwinian emphasis on individual advantage with the apparently cooperative behavior of collectives thus lay at the heart of much evolutionary theorizing in the 1960s and 1970s. It is therefore not surprising that the prisoner's dilemma game (and game theory more generally) emerged in biology at precisely this time in association with debates over the evolutionary origins of altruism and aggression in humans and non-humans alike.

A second development that helped bring the prisoner's dilemma matrix into the natural world involved specifying more precisely the nature of the "payoffs" at stake in a game. Humans play games for money and pleasure; to generalize to other life forms, neo-Darwinian evolutionary theorists would need to find new metrics for the costs and benefits arising from evolutionary adaptations—in effect, a "utility function" for life itself. The value of evolutionary adaptations would not be measured by their benefit to species or even to individual organisms; they were preserved if they helped to perpetuate the genes that controlled them. This perspective was pioneered by the British biologist William D. Hamilton in several papers published in 1963 and 1964 that created the theory of "kin selection." Hamilton directly tackled the problem of how to explain altruistic behavior in terms of its evolutionary advantage to individuals. Organisms did not dispense altruism out of concern with collective solidarity; rather, altruistic behavior evolved if its benefit to an individual organism's inclusive fitness—a quantity that included the survival benefit to other organisms weighted by their degree of genetic relatedness—outweighed the costs to the individual.[49]

It was in this context that Hamilton first introduced ideas from game theory into his work in a 1967 paper that sought to explain the evolution of the sex ratio, especially the existence of highly lopsided sex ratios in a number of species of insects. Sex ratios did not emerge to maximize the reproductive success of the species as a whole; rather, to a greater or lesser degree of realism, organisms behaved as if they played games with each other in which the sex ratio in their offspring represented their "strategy" in the game. Since natural selection favored individuals with greater fitness relative to the rest of the population, Hamilton argued that organisms would evolve as if they selected two-person zero-sum game theory's "minimax strategy"—or as Hamilton put it, an "unbeatable" strategy against which none other could do better. Depending on the structure of the population in question (e.g., the degree of reproductive mixing that its ecology permitted), species would develop different sex ratios.[50]

Organisms thus had the interests and competitive drive to be game players. But while Hamilton could use the language of "games," "strategy," and "choice" in talking about the behavior of tiny wasps and mites, he was clearly uncomfortable about the implications of this language. Some organisms certainly seemed to be capable of sensing their environments and adjusting their behavior accordingly. At the same time, imputing intentionality or the conscious ability to respond to environmental stimuli seemed not far from attributing to species precisely the kind of "collective interests" that neo-Darwinian theory sought to banish. Indeed, as Hamilton would remark in a 1968 letter to his friend and colleague George Price, while "with the tiny animals discussed I think it extremely unlikely that they are able to play the suggested 'game' intelligently, or recognize their own 'sex-ratio types,'" nevertheless "I think there is an interesting theoretical problem as to how sex ratio behaviour should be expected to evolve if an intelligent animal like man was to find itself in the situation described."[51]

In this regard Hamilton was particularly intrigued by Anatol Rapoport's analysis of the prisoner's dilemma game, which seemed to highlight the divergence between the behavior of animals with "exceptional intelligence," like humans, and those without. By 1967, as efforts to develop an intellectual framework for evaluating arms control schemes continued to focus on prisoner's dilemma and related games, Rapoport could suggest that collective and individual rationality in prisoner's dilemma could be reconciled if one considered the problem within a broader framework of "metalogic," where players could consider strategies conditional upon the different possibilities for how other players in the game might act.[52] However, Hamilton suggested, "against this the model in my paper seems to show that the proposed 'solution' does not hold for the animals discussed under natural selection, and I am doubtful whether intelligence makes much difference to the kind of solution that is possible. I am sure that prisoner's-dilemma situations are common and important in biological evolution."[53] More broadly, Hamilton mused on whether the evolution of reasoning abilities (in the form of speech, memory, and cognition) would do much to resolve a prisoner's dilemma–type paradox. Perhaps the ability to communicate would also bring with it the ability to lie? And perhaps deceit would only be enhanced in a cognitive arms race, as organisms evolved ever subtler and more complex ways to deceive one another? In the end, he would conclude that culture and altruistic values simply formed a "higher hypocrisy" intended to fool our fellow human beings.[54] There seemed to be no way to bridge the selfishness of genes with the kind of social cooperation envisioned by Rapoport and others approaching cooperation from a psychological perspective.

The perfect opportunity for Hamilton to share his musings on this problem arose in May of 1969 when he was invited to participate in a major interdisciplinary conference, held at the Smithsonian Institution in Washington DC, that sought to explore the lessons of recent developments in ethology and the study of animal behavior for understanding the problem of human violence on both the national and international levels. Naturally, Hamilton's contribution to the conference provided little solace for those who hoped to find a biological basis for altruism on either side of the human/nonhuman divide. The basis for his pessimism was a model of the possible pairwise interactions between individuals in which two genetic strategies are possible: one associated with a "normal" gene (N), and the other with a mutant "selfish" gene (M), with gene frequencies p and q, respectively. Any given type of interaction could be represented by a 2×2-game matrix, with "payoffs" denominated in abstract units of evolutionary fitness (figure 5.4).

However, instead of trying to solve this game for the probabilities with which the players would adopt the two strategies (as a game theorist interested in human behavior might have done), Hamilton instead computed a difference equation that described how these "probabilities"—in this situation, the gene frequencies p and q—would evolve. If the game payoffs are those of a prisoner's dilemma game, and even if the selfish gene is initially rare, Hamilton's difference equation suggested that it would spread through the population over time. Hamilton saw in this result a fundamental lesson of evolutionary biology: it is not as important to choose an

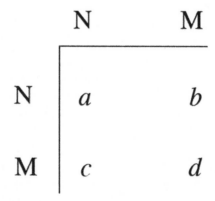

Figure 5.4. The outlines of Hamilton's PD matrix; a population of "normal" genes (N) is invaded by a "mutant" selfish gene (M). (W. D. Hamilton, "Selection of Selfish and Altruistic Behavior in Some Extreme Models," in *Man and Beast: Comparative Social Behavior*, ed. J. F. Eisenberg and Wilton S. Dillon [Washington DC: Smithsonian Institution Press, 1971], 62.)

"optimal" strategy as it is to choose a strategy that is simply better than those employed by other organisms in the same population. From an evolutionary perspective, it does not pay to cooperate in prisoner's dilemma situations.[55]

While the result seemed decisive, its apparent contradiction of Rapoport's results nevertheless bothered Hamilton—despite the fact that Rapoport's results emerged from considerations of what we might imagine to be distinctively human behavior. Hence the balance of the paper reads as a long deliberation on the complicated and unsettled relationship between game theory as it appeared in psychology on the one hand and Hamilton's vision for evolutionary theory other the other. Hamilton recognized that the theory of games current in the former fields "presupposes being able to think, and, potentially, to communicate." This cognitive-psychological aspect of game theory made the theory a convincing model of human conflict and cooperation, but it also led to the recognition of irreconcilable conflicts between individual and group interests—interests that he felt simply did not have clear counterparts in nonhuman populations, since animals did not possess the same subtleties of communication and cognition found in humans. Perhaps games needed to be solved in different ways for humans and nonhumans, with something approximating Rapoport's "cooperative" solution for humans and Hamilton's "selfish" solution for nonhumans. In such a case, Hamilton's solution would have little relevance to the human problem of "how it is rational to act" when thrust into a prisoner's dilemma, despite the fact that natural selection "has made us almost all that we are."[56]

Two encounters changed Hamilton's outlook over the next decade. The first was a chance meeting at the Washington conference between Hamilton and Robert Trivers, then a graduate student working with Harvard primatologist and anthropologist Irven DeVore. Trivers had begun investigating the anthropological literature on "reciprocation"—the exchange of favors and aid—in hopes of explaining the emergence of altruistic behavior among primates and among nonhuman organisms more generally. Hamilton's paper at the Washington conference came as a revelation to Trivers. Apparently, in his presentation at the conference, Hamilton added a coda to his paper in which he gestured to existing work on repeated prisoner's dilemma games and suggested that a repeated-game framework might demonstrate the feasibility of cooperation as a solution to Hamilton's prisoner's dilemma game. Trivers quickly saw the relevance of such games to his problem of reciprocated altruism.[57] Shortly thereafter, in his groundbreaking paper, "The Evolution of Reciprocal Altruism," he noted "the relationship between two individuals repeatedly exposed to symmetrical reciprocal situations is exactly analogous to what game theorists call the Prisoner's Dilemma. . . . Iterated games played between the same two individuals permit each player to respond to the behavior of the other." The extent of the communication going on in Trivers's models was questionable, given that his mathematical formulations did not require individual organisms to remember their interactions with others or modify their behavior in the present in response to their expectations about the future. Indeed, this was critical to Trivers's argument since, following Hamilton's theory of kin selection, his goal was to "take the altruism out of altruism," to remove the necessity for altruists to have intentions or choice. However, drawing on calculations suggested by Hamilton, he was able to show that in populations of organisms where players would interact a given number n times, it was possible for altruistic genes to maintain themselves against invasion from selfish genes, even if they might not spread from a low initial probability. Trivers proceeded to identify numerous examples of reciprocal altruism: mutualistic grooming, altruistic alarm calls in birds that alert the flock to the approach of predators, and so forth.[58]

A second encounter that changed Hamilton's mind about the possibility for cooperation evolving in repeated prisoner's dilemma games came when he left Imperial College for the University of Michigan. By this point, both Anatol Rapoport and the Center for Research on Conflict Resolution had long since moved to other institutions. However, not long after his arrival in Ann Arbor, Hamilton became acquainted with the political scientist Robert Axelrod, who was then designing a series of computerized "tournaments" between strategies for repeated two-player prisoner's dilemma games submitted by game theorists across the country. The reigning champion strategy—submitted by none other than Rapoport himself—would prove to be one of "tit for tat" in which one player would reward the other's cooperation in one move with cooperation in the next (and likewise, punish noncooperation with noncooperation).[59] Axelrod's style of investigation must have appealed to Hamilton, especially since the computerization of strategies could banish unrealistic assumptions about the kind of interactions the players could have, reducing appeals to the memory, cognition, or reasoning ability of the fully mechanized game "players." One product of Hamilton and Axelrod's Ann Arbor meeting was a landmark 1981 paper, "The Evolution of Cooperation," which demonstrated that tit-for-tat solutions to prisoner's dilemma could be found if the chance of the

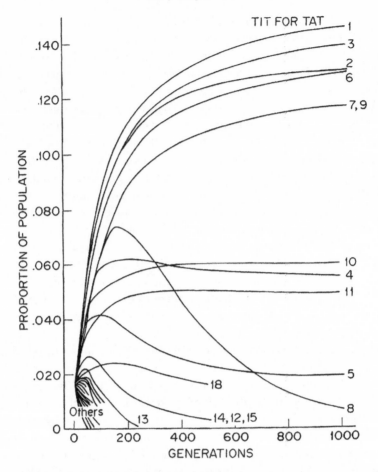

Figure 5.5. The evolution of "Tit for Tat." (Robert Axelrod, "More Effective Choice in the Prisoner's Dilemma," *Journal of Conflict Resolution* 24, no. 3 [1980]: 379-403.)

players meeting again was high enough (figure 5.5). The article would later provide the title for Axelrod's 1984 collected essays on cooperation in prisoner's dilemma games, *The Evolution of Cooperation.*[60]

It is interesting to note that in the opening paragraphs of *The Evolution of Cooperation*, Axelrod invokes Hobbes's famous description of life in the state of nature that existed prior to the establishment of governments: "solitary, poor, nasty, brutish, and short." Clearly Axelrod saw repeated games and reciprocation as a way of explaining the emergence of *human* institutions, social norms, and morality—concepts the evolutionary biologists had little use for. Moreover, Axelrod had no problem granting human game players the ability to pursue goal-directed deliberate action, the neo-Darwinians' bête noire. But his starting point was nevertheless the state of nature and its automaton inhabitants that Hamilton and others insisted upon as the basis for their evolutionary models. Even as reciprocation and repeated prisoner's dilemma games would again receive significant attention within both economics and political science in the 1980s in the wake of Axelrod's work, this legacy of game theory's sojourn in evolutionary biology would linger. Cooperation in the prisoner's dilemma did not need to emerge from conscious reasoning or rational calculation, but could simply result from the evolutionary dynamics of selfish genes.

5.4. Writing Matrices, Locating Rationality

From mathematical optimization problems to the study of insect populations to reflections on the Cold War arms race itself, game theory's characteristic matrix notation helped intellectuals of this period to find prisoner's dilemmas almost everywhere. Examinations of reasoning, rationality, and choice were no longer totally bound to the specifics of particular situations, but could jump from one setting to another in tandem with transfers of research programs, investigatory techniques, careers, and insights. And throughout, the "ideal type" of Cold War rationality could move outward from the specific contexts that had nurtured its component parts to gain consideration from individuals spread out across a much broader intellectual landscape. Most notably, the rule-following computer appears throughout this history as a point of reference for exploring rational conduct, even as it simultaneously suggested the fundamental unreasonableness of Cold War rationality. The perfect Cold War rational agent—whether embodied in John Nash's ideal experimental game-playing subjects or in Axelrod and Hamilton's simulated game-playing organisms—may have possessed prodigious calculational abilities, but he frequently also lost a set of faculties and qualities classically associated with reasoning: memory, choice, consciousness, goal-directedness, and even intelligence. This loss was alternately derided (for instance, by the social psychologists) or embraced (by the evolutionary biologists) in particular contexts and for particular reasons, yet it remained a central feature of prisoner's dilemma–inspired arguments about the nature of rationality until at least the 1980s.

The apparent reconciliation of self-interest and cooperation in prisoner's dilemma via "tit for tat" as well as a number of related mathematical results preceded the end of the Cold War by only a few years. Just how connected these two developments were is open to debate. Certainly Axelrod's work gained a wide audience in political science and the arms control community, and "reciprocation" and "reciprocity" dovetailed with buzz-words common even among top-level policymakers since the advent of concerted arms limitation talks in the 1970s. Yet the central results of evo-lutionary game theory suggested that a substantively rational outcome to the prisoner's dilemma might have less to do with the intelligence or deci-sion-making abilities of the players (or the theorem-proving ability of their policy advisors) and more with the blind historical processes that guided them. As with Adam Smith's "invisible hand," which ensured productive economic behavior without requiring good intentions on the part of the butcher or the baker, a cooperative conclusion to the Cold War need not have emerged from actors who were reasonable in any meaningful sense.

And as cognitive psychologists came to assert with increasing volume throughout the 1970s and 1980s, this was probably for the best. In the eyes of many, humans in this era fell far short of the standards presumed by Cold War rationality. This growing recognition, emerging in tandem with the breakup of the Soviet Union and the dissolution of the national security consensus in the United States, ultimately brought about the privi-leging of "irrationality" as the prime characteristic of human decision mak-ing and the splintering of Cold War rationality.

SIX

The Collapse of Cold War Rationality

How would the Soviets react this time?[1] Following price increases dictated by the Warsaw government in July 1980, Polish citizens protested and went on strike. In August, the electrician Lech Wałęsa led sixteen thousand workers in the occupation of the Lenin shipyard in Gdansk. Strikes at other shipyards followed. On September 6, Edward Gierek—leader of the Polish Labor Party since 1970—was replaced by the reformer Stanisław Kania. On November 10, the high court declared the new independent trade union Solidarity (*Solidarność*), led by Wałęsa, to be legal. Within less than two years, some nine to ten million Polish citizens had become members, approximately a quarter of the whole population. On February 9, 1981, General Wojciech Jaruzelski became head of government. He officially agreed with Kania concerning the reform course but also indicated that he would fight against unrest. In the following months, the Polish Labor Party began to accept devout Christians as members, secret elections were held at a party convention, censorship was liberalized, yet strikes and demonstrations continued.

From Moscow's point of view, things could not go on like this. The Soviet Union demanded a purge of the Labor Party, and initiated military field exercises on the Polish border. The message was clear, given the Soviet invasions of East Germany in 1953, Hungary in 1956, Czechoslovakia in 1968, and the Brezhnev doctrine of limited sovereignty of Communist countries. Kania was ousted as party leader and, on October 18, succeeded by Jaruzelski. On December 13, the now sovereign Military Council for National Salvation imposed martial law throughout the country. Still, even though Wałęsa was detained along with several thousand other Solidarity members, strikes and unrest continued throughout 1982 (figure 6.1).

The United States and its NATO allies were monitoring the situation

Figure 6.1. Cover of *Time*, December 29, 1980.

closely too. President Ronald Reagan blamed the Polish government and the Soviet Union for the escalation and imposed economic sanctions upon both countries. In Western Europe, large demonstrations were organized in support of Solidarność. To be sure, NATO countries had not resorted to military force during earlier Soviet interventions in Warsaw Pact countries. Because of the declaration of martial law, a Soviet intervention had also so far been avoided. But the Polish situation further complicated already vexed East–West relations. The Soviet Union had invaded Afghanistan in 1979. In the same year, NATO made its Double Track Decision, which dictated that either the Soviets would have to withdraw their new SS-20 missiles from Central Europe, or NATO would install its own intermediate-range nuclear weapons. Negotiations had been under way since 1981 but had stalled. Reagan was reluctant to conduct arms control negotiations with the Soviets. In 1982, the United States and the Soviet Union began the Strategic Arms Reduction Talks, but Reagan also increased spending for the military to record levels and, in March 1983, announced his Strategic Defense Initiative for missile defense. Especially to many Europeans, the whole situation looked like a return to the Cold War after the years of détente.

So how would the Soviets react? Would Poland be invaded? Would relations with NATO be frozen? Could it lead to an escalation of the Cold War? Could experts and politicians make reasonable predictions and decide rationally what to do in a situation fraught with uncertainties and risks? In this chapter, we look at an approach in research on human rationality that became strong in the late 1970s and '80s, the so-called heuristics-and-biases approach, the results of which were seen by many as supporting the claim that humans—both experts and laypersons—are often highly irrational judgment and decision makers. This approach became directly applied in many areas of political science and policy analysis, from voting behavior to arms negotiations to international conflict. However, starting in the late 1970s, philosophers and psychologists raised strong criticisms of the heuristics-and-biases approach, leading to a fragmentation of the very concept of rationality. These "rationality wars" have frequently been neglected by political scientists. They should not. It is now regarded as a truism that in 1981 or 1983 it was hard to recognize not only the risks but also the new opportunities related to the Polish crisis—that the Soviet system would collapse and the Cold War end within a few years. A deeper lesson, however, may be that we do not possess an account of rationality on which the relevant political expertise could be grounded. The Cold War hope for a unified concept of rationality providing the tools for perfectly understanding, predicting, and dealing with political conflict perished.

6.1. Psychology of Reasoning: Two Domains

Let us begin with a typical psychological experiment. In July 1982, Amos Tversky and Daniel Kahneman presented various hypothetical scenarios to 115 participants at the Second International Conference of Forecasting in Istanbul, Turkey, asking them for their comparative probability. A first group had to consider the case of "a complete suspension of diplomatic relations between the USA and the Soviet Union, sometime in 1983"; a second group was confronted with a scenario of "a Russian invasion of Poland, and a complete suspension of diplomatic relations between the USA and the Soviet Union, sometime in 1983." This latter scenario was judged to be more probable than the former (geometric means: 47 percent and 14 percent, respectively). Was this the rational answer? From the point of view of mathematical probability theory, the answer is "no." The latter statement could not be more probable than the former, because the intersection of two events, "invasion and suspension," cannot be more probable than one event, "suspension."[2]

Since the 1960s, the psychology of human rationality had become an increasingly popular research field. Researchers studied how well humans reasoned about abstract as well as concrete problems, the conditions for and limits of human rationality, and the reliability of reasoning by both laypeople and experts. Many psychologists concluded that the results were sobering, even alarming. Moreover, their views were adopted by political scientists and policy analysts. This chapter examines psychological research on rationality and its applications in the final decade of the Cold War, as well as criticisms of the program by philosophers and psychologists—critiques that to date have had remarkably little effect upon the branches of political science and policy analysis that seek psychological underpinnings of political judgment and decision making.

Two main areas of this research program on the psychology of rationality can be distinguished: on the one hand, reasoning tasks involving deductive or logical thinking; on the other, judgment and decision making under conditions of uncertainty. Both areas are related in goals and methods, but we place particular emphasis on the second, because it was often applied to politics and international relations, clearly urgent arenas of judgment under uncertainty.

These psychologists understood human rationality as the ability to reason in accordance with norms of logic, probability theory, statistics, and formal decision theory. This conception of rationality was in full accord with the formal theories of reasoning we have encountered in earlier chapters. Some such normative standards were also necessary for psychological experimentation about rationality, since otherwise it would be unclear how one would be able to say that subjects got it right or wrong, reasoned well or reasoned badly. Moreover, instructing subjects on how to reason better also presupposed standards. However, who could legitimately set those standards in the first place? Many psychologists took the basic rules of logic and probability theory to be more or less uncontroversial. They also sometimes assumed that they could rely on the claims of logicians that all of propositional logic could be derived from the imperative to avoid contradictions, that probability theory could be given a Dutch book justification (that is, an argument showing that if one does not accept the laws of probability as conditions for the coherence of one's beliefs, one is bound to accept bets one is certain to lose); and that analogous claims could be made for the theory of games and decisions.[3]

The work by the British psychologist Peter C. Wason (1924–2003) became the model for most empirical studies of deductive reasoning. Wason, who spent most of his career at University College London, published widely on thinking and reasoning and was a passionate player of, author about, and international master in correspondence chess. Research on reasoning attracted him "because most things in life seemed unreasonable."[4] Wason had a talent for inventing unusual studies. Not all of his ideas were successful. For instance, he once investigated the hypothesis that subjects who had to write in invisible ink would write more fluently. The editor of a journal was willing to publish it—but, as the story goes, only in invisible ink.[5] In any event, most important among his reasoning studies became known as the "four card test," or the "Wason selection task" (WST). This is the grandmother of all reasoning tests, invented in the early 1960s and first published in 1966. This test has since been deployed and discussed in numerous variations.[6] We shall examine it more closely below.

The second group of reasoning studies followed in many respects the basic methodology of the WST, but extended it to probability, uncertainty, or risky decision making. In this realm, the heuristics-and-biases approach, developed especially by Daniel Kahneman (b. 1934) and Amos Tversky (1937–1996) during the 1970s and '80s, proved the most influential. Both had worked in the Israeli army: Tversky as an officer, Kahneman as a psychologist. Kahneman integrated his research with his experience of testing recruits' and instructors' abilities for the military, which emphasized avoiding errors in judgments. As he himself reported, some of the test results directly led to claims about judgmental biases, or "illusions." Starting in 1969, this focus on mistakes was integrated into Tversky's behavioral decision research.[7] He had grown skeptical of the claims of statisticians such as Leonard J. Savage (1917–1971) or psychologists such as Ward Edwards (1927–2005), who maintained that people by and large obeyed the rules of Bayesian statistics and expected utility theory.[8] For instance, Kahneman was involved in assessments of candidates for officer training. Subjects were asked to perform stressful tasks without wearing any signs of their rank—they were identified merely by numbers (similar to some of the experiments described in chapter 4). The experimenters had to assess leadership quality in terms of ratings of candidates' potentials, and were quite convinced of these predictions. On "statistics day," however, the psychologists got feedback from the officer training school, and it turned out that the assessments did not match with the feedback. The psychologists could not really tell "who would be a good leader and who would not."[9] Kahneman dubbed this the "illusion of validity," and it entered an influential article he later on published with Tversky in 1973.[10]

Kahneman and Tversky subsequently held positions at several universities. Most of their careers they both worked at the Hebrew University in Jerusalem and then, starting in 1978, at Stanford University. They jointly published (sometimes with additional collaborators) a large number of articles and several books, including much-cited papers in the journals *Science* and *Econometrica*. Their heuristics-and-biases approach found applications outside of psychology, often with its inventors' encouragement. Tversky was a cofounder, with Kenneth J. Arrow and others, of the Stanford Center on Conflict and Negotiation. Furthermore, the heuristics-and-biases approach fertilized economic theories that sought to construct more realistic models of economic judgments, decisions, and behavior patterns. Kahneman received the Nobel Prize in Economics in 2002, the only other cognitive scientist so honored since Herbert Simon in 1978. Tversky died in 1996, but Kahneman emphasized that the prize was given for work that resulted from their collaboration, especially for "prospect theory"—an alternative to classical utility maximization for decisions under uncertainty (see section 6.5).[11]

Let us now see how these psychological studies on reasoning worked, and how they afterward came to be applied to political expertise and events.

6.2. Wason's Four Card Test

A popular theory attracts critics as well as adherents. The Swiss psychologist Jean Piaget (1896–1980) developed a celebrated stage-model of human cognitive development: the sensorimotoric stage (0–18 months); the preoperational stage (18 months–7 years); the concrete operational stage (7–12 years); and, most important for our present topic, the formal-operational stage (from 12 years onward). He supposed that, beginning in this last stage, normal humans master typical logical operations, and that this is constitutive of rationality: "reasoning is nothing more than the calculus embodied in the propositional operations."[12] True or false?

The famous—and notorious—Wason four card test was supposed to answer this question. It looks deceptively simple at first glance but leads to all sorts of puzzles upon closer examination. In a typical version, subjects were presented with four cards, of which they saw the front side only. Two of the cards showed letters (D or E), and the other two numbers (3 or 4) on their visible sides. Cards with letters on their front sides were marked with numbers on their back sides, and vice versa—which numbers or letters was unknown to the subjects. Subjects were then asked to evaluate statements about these cards.

Subjects mostly picked out only the E-cards, or the E-and the 4-cards. Fewer than 10 percent of the test subjects chose the correct solution: the E-and the 7-cards.[13] At least, this is the correct solution according to standard propositional logic. This logic views conditional statements of the form "If p, then q" as *material implications* (usually represented as $p \rightarrow q$). Such a statement is considered false if, and only if, p is true and q is false. In all other cases the statement is regarded as true. For instance, if you assert that "if Dick comes to the barbeque, then Daisy also comes," the whole statement is obviously false if Dick comes but Daisy does not show up. Similarly, in the WST, the subject should turn over the E- and 7-cards, and only these ($=p$ and not q). There's no need to turn over the T-card; a material implication is already true if its antecedent (the p part) is false, and "T" represents just that (figure 6.2).

The good news was that subjects indeed rarely picked the T-card, but this did not catch psychologists' attention. They were captivated by the bad news: subjects thought it was a really good idea to pick the even-numbered

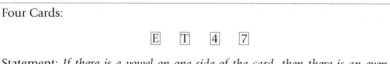

Four Cards:

E   T   4   7

Statement: *If there is a vowel on one side of the card, then there is an even number on the other side.*

Question: What card(s) has (have) to be turned over to see if this statement has been violated?

Figure 6.2. The Wason selection task. (Adapted from Wason 1968.)

4-card. Assessed by the standards of propositional logic, it is unnecessary to check this card because an implication is true if the consequent (the *q* part) is true.

Wason originally proposed two possible explanations: The first option was that subjects were denying that "if *p*, then *q*" statements with false antecedents have any truth value at all and instead considered such statements "irrelevant."[14] It was as if, as he later remarked, they were following the critics of ancient Stoic and Megarian schools.[15] (When one says "if it is sunny, then I'll go swimming," and if it actually rains, most people would think it odd to infer from the rainy condition that the whole statement must be true.) However, Wason preferred another explanation: there is a "need to establish the 'truth' of the statement," which "predominates over the instruction." Subjects had been explicitly instructed to look for the cards that show whether the conditional statement had been *violated* but did not follow the instruction. Wason described this as a "bias toward verification" and attributed it to the "tendency to confirm, rather than eliminate hypotheses" in science.[16] It was as if the subjects were de facto rejecting Karl Popper's falsificationist philosophy of science.[17] In any case, for researchers who accepted the results of the WST, the test refuted the Piagetian idea that normal human adults are able correctly to apply standard rules of propositional logic.

6.3. Reasoning under Uncertainty: Heuristics and Biases

The multiplication of biases in psychology intensified when Kahneman and Tversky entered the scene around 1970. Alongside collaborators and other research groups,[18] they developed many experiments dealing with judgment and decision making concerning matters of probability, statistics, and rational choice. These studies won a wider audience than logical tests, perhaps because in real life purely logical rules rarely prove sufficient. Moreover, uncertainties and risks are everywhere: in gambling and investing money most obviously but also in driving cars, making health decisions, or forecasting political developments. As in the case of the WST, however, the typical experiments were often remote from real-life affairs, bearing a closer resemblance to classroom exams.

In one highly influential study (financially supported, like many other publications of the program, by the US Office of Naval Research), Tversky and Kahneman tested the ability of subjects to follow a basic rule of mathematical probability theory: the conjunction rule.[19] It states that an event *A* can never be less probable than a conjunction of the (independent) events

A and B: Prob $(A) \geqslant$ Prob $(A \& B)$. Subjects' adherence to this rule was tested by the "Linda problem." Here is a simple version of the task:

> Linda is 31 years old, single, outspoken and very bright. She majored in philosophy. As a student, she was deeply concerned with issues of discrimination, social justice, and also participated in antinuclear demonstrations. Which statement is more probable?
>
> (T) Linda is a bank teller.
> (T&F) Linda is a bank teller and active in the feminist movement.

About 85 percent of 142 subjects chose the answer "T&F" and thus violated the conjunction rule.[20] As in the case of the WST results, therefore, things were not looking good for human rationality. At the same time, it must be pointed out that a certain notion of rationality was assumed to lead to this conclusion: one in which formal rules of logic and probability theory are as essential as the content of the reasoning task is taken to be inessential. In the statement (T&F), "and" has to be understood as the logical AND, such that the two conjuncts are seen as independent events. Also, the description of Linda is strictly irrelevant to the question of whether (T) or (T&F) is the right answer.

The test for the conjunction rule was subsequently varied in many ways. Different cases of persons, events, and the like were described and connected to various questions—including the study about the scenarios concerning a possible Soviet intervention and US–Soviet relations mentioned in section 6.1. But even when subjects were told the correct answer, the number of incorrect answers barely decreased. Kahneman and Tversky contended that the results in the Linda problem were stable, and that they were evidence of a stubborn and systematic incompetence in human reasoners.[21] They explained this "cognitive illusion"[22] by saying that subjects err systematically because of what they called the "representativeness" of certain features of the description of Linda: If someone is concerned with issues about discrimination, social justice, and nuclear weapons, and is single, many people think that these properties are likely to be representative for being a feminist as well.

In addition to representativeness, two other major heuristics invoked by Kahneman and Tversky were "anchoring and adjustment" on the one hand (the tendency to generate estimates by taking a value suggested by some information given and then adjusting it upward or downward), and on the other, "availability" (the tendency to predict the frequency of an event, or a proportion within a population, depending on how easily an

Figure 6.3. "Hindsight Bias." (Cartoon by Peter Nicholson,
The Australian, October 16, 2002.)

example can be brought to mind). Representativeness also allegedly impelled people systematically to ignore base rates when making diagnoses. According to one study, even statistically educated medical staff members at Harvard Medical School committed such base rate fallacies when being asked to estimate the probability of breast cancer given certain symptoms.[23] Subjects also misperceived randomness, overestimated their abilities as car drivers, exhibited exaggerated levels of confidence in their judgments about general-knowledge questions ("overconfidence"), falsely believed in retrospect that they had known the right answer all along ("hindsight bias"; see figure 6.3), and could not, would not calculate expected utilities according to economic decision theory.[24]

6.4. The Irrationality Message

Such psychological studies on reasoning seemed to show that normal adults do not follow the canons of formal logic, probability theory, statistics, or decision theory. Instead, so the story went, humans apply certain "heuristics": rules of thumb that work fairly well in some contexts but which are not generally valid and which lead to "biases" (systematically incorrect results). What is more, it was often inferred that the biases were systematic and ineradicable, thus warranting a further inference to pervasive, incorrigible human irrationality. Kahneman and Tversky did not explicitly draw this latter conclusion. They emphasized that a focus on errors and fallacies was necessary for methodological reasons: just as in perceptual theory, in order to figure out how the mind works it is useful to look at what occurs when things go wrong, as in the case of perceptual illusions, which are also systematic and ineradicable.

But they did draw very similar inferences. With respect to the error called the "law of small numbers"—the tendency to draw conclusions from very small samples rather than heeding the rule that only large samples will be representative of the population from which they are drawn—they wrote: "The true believer in the law of small numbers commits his multitude of sins against the logic of statistical inference in good faith . . . His intuitive expectations are governed by a consistent misperception of the world."[25] In another study on decision theory, they questioned the "favored position" of the "assumption of rationality" in economics, allegedly treated by economists as "a self-evident truth, a reasonable idealization, a tautology, a null hypothesis." It was, they claimed, an assumption impervious to empirical falsification:

> the assumption of rationality is protected by a formidable set of defenses in the form of bolstering assumptions that restrict the significance of any observed violation of the model. In particular, it is commonly assumed that substantial violations of the standard model are (i) restricted to insignificant choice problems, (ii) quickly eliminated by learning, or (iii) irrelevant to economics because of the corrective function of market forces.[26]

Other researchers did assert outright that the reasoning studies, whether on logical or on statistical topics, had "bleak implications for human rationality"[27] or concluded that "the selection task reflects [a tendency toward irrationality in argument] to the extent that subjects get it wrong. . . . It could be argued that irrationality rather than rationality is the norm."[28] Psychologist Massimo Piatelli-Palmarini (then a research scientist at MIT, later professor in Milan) announced that the studies support the following "probabilistic law: Any probabilistic intuition by anyone not specifically tutored in probability calculus has a greater than 50 percent chance of being wrong."[29] Or, in a similarly pessimistic vein: "One might draw rather cynical conclusions. . . . Human reasoning is fundamentally flawed."[30] Only bad news is good news.

6.5. Political Applications of the Heuristics-and-Biases Program

Thus, the heuristics-and-biases research program reinforced the age-old view that humans are susceptible to follies and fallacies. The sweeping claims about human irrationality won this research a wide audience, not only in other sciences such as in economics and medicine but also in more popular books and in media accounts.[31] Most important for the present study, it was also applied to political science and policy analysis, opening up a space for new questions: what explains human judgments and decisions in times of international crisis? Moreover and more urgently, can we trust politicians' and political experts' rationality? If their judgments and decisions were irrational, why was that so? Could anything be done about it?

Biases and heuristics were invoked to study political actors and experts, voters and politicians alike, in areas such as exaggerated risk estimations concerning technology and international conflicts,[32] the "illusion" that an individual's vote matters in large elections,[33] or the unwillingness to correct one's political beliefs about the past and the future in light of evidence.[34] One RAND study also attempted to integrate heuristics-and-biases psychology with the theory of "groupthink" analyzed in chapter 3.[35] Even the behavior of nation-states, especially in times of international crises and conflict, was studied this way: for instance, concerning Japan's decision to attack the United States in 1941,[36] the Suez crisis of 1956, the Cuban Missile Crisis,[37] nuclear deterrence more generally,[38] or the Iran hostage rescue mission.[39] The Polish crisis and the arms race during the 1980s that opened this chapter were not exempt from this approach. Contemporary analysts described Wałęsa's confrontational strategy against the Polish government as one cause of the imposition of martial law (something the leadership of Solidarity had not intended), and so as a clear case of "overconfidence." Also, Western countries were reminded that détente would reduce opportunities for conflict, help to integrate Communist countries into the world economy and so put a "premium on rational calculation."[40]

Let us consider a few more detailed studies applying heuristics-and-biases psychology to such policy topics. In her early work, cognitive scientist Nancy Kanwisher, recipient of a MacArthur Foundation Fellowship in Peace and International Security in 1986, applied heuristics and biases to foreign policy matters. She argued that many tenets in American security policy were "held with strong but unwarranted conviction,"[41] pointing to Tversky and Kahneman's aforementioned study,[42] in which various hypothetical scenarios had been presented to forecasters and planners at the height of the Polish crisis. She drew several lessons from their disappointing performance on the psychologists' tests. First, experts might not be very reliable. Second, in explicit opposition to Herman Kahn's penchant for hypothetical worst-case scenarios, Kanwisher called attention to the range of possibilities that must be thought through in strategic analysis. Kahn had warned against using such scenarios as tools for *prediction*, but this "reflected some psychological naiveté" since people would still often understand the scenarios in a predictive manner.[43] Moreover, she pointed to evidence according to which the overwhelming majority—allegedly 90 percent—of RAND studies used extreme scenarios of a Soviet surprise attack, yet when Kahn and Thomas Schelling had asked their collaborators about the probability of such an out-of-the-blue attack, this scenario was estimated to be the least likely. Thus far, Kahneman and Tversky's message that humans cannot follow standard formal norms of rationality was far from being confirmed. But Kanwisher was unimpressed by this point: she countered that the massive favoring of worst-case scenarios had the practical effect of not investing much into research on more probable developments.

Kanwisher also exploited psychological studies on overconfidence, on the "homeopathy heuristic"—the tendency to think that effects must resemble their causes (e.g., big effects must have big causes)—or on the "illusion of control."[44] She used these tools simultaneously to explain and to evaluate critically the logic of nuclear deterrence, particularly the tenet that escalation can be controlled by enlarging one's options, whereas in fact one might be sliding uncontrollably toward war.

Others analyzed the arms race and stalled disarmament negotiations from the perspective of the heuristics-and-biases program. Take the prominent account of decision making emerging from Kahneman and Tversky's approach, known as "prospect theory." This tries to "repair" expected utility theory (understood as a descriptive theory) by adding parameters meant to capture risk aversion, loss aversion, and other psychological factors, while retaining the notion that humans still algorithmically compute losses and gains, albeit unconsciously.[45] Tversky himself, with his collaborator George Quattrone, claimed that prospect theory could be used to understand why in "negotiating over missiles . . . each superpower may sense a greater loss in security in dismantling of its own missiles than it senses a gain in security from a comparable reduction made by the other side."[46]

Reagan's reluctant approach to arms control negotiations in his first term could also be explained differently within the heuristics-and-biases

approach—for instance, as a bias due to the availability heuristic. The political scientist Rose McDermott, one of the most fervent advocates of using heuristics as explanatory tools in policy analysis, claimed that Reagan
had not rationally calculated the probabilities of future Soviet expansion
nor coolly assessed Soviet noncompliance with arms control agreements.
Instead, McDermott argued, he reverted to his deeply held convictions that
the Soviet system was fundamentally illegitimate, that the Soviet Union
constantly aimed at expansion, that appeasement would lead to war, that
military strength would conversely secure peace, that mutual assured destruction (MAD; see chapter 3) was not a good strategy, that the Strategic Defense Initiative (SDI) might provide an additional tool for deterring
the Soviets, and so on. Such beliefs created a schema of "what is salient
and available to a decision maker when formulating policies"[47]—alas, a
biased and therefore irrational schema. As McDermott claimed, "Ambiguous as well as contradictory information is interpreted in such a way as
to bolster existing theories about the world, while contradictory evidence
is either ignored or discounted in order to maintain preestablished belief
systems."[48] Discussing then secretary of defense Dick Cheney's attitudes in
the late 1980s, diplomatic historian Richard Immerman reasoned similarly
that "cognitive psychologists uniformly agree that once we have formed a
belief we are reluctant to discard or even qualify it. . . . No matter how
forcefully Mikhail Gorbachev pushes for arms control, he will not alter . . .
Cheney's estimate of the Soviet threat. . . . Cheney's core beliefs and images
remain unshakeable."[49] Whichever theoretical notions used to account for
the reluctance of the United States toward arms control during that time,
analysts frequently viewed political decision makers as biased by certain
heuristics instead of following formal rules of classical models of rational
choice or rational belief formation.

6.6. Irrationality, Schmirrationality: Dissenting Voices

Connecting political science with what was conceived of as cutting-edge
cognitive science exuded a certain interdisciplinary glamour. Policy analysts and social scientists were happily applying heuristics and biases in order to understand and tame human conflict, and thus appeared to use the
most recent theories from cognitive science. However, since the late 1970s,
critics from philosophy and psychology started to worry about the empirical soundness and conceptual and methodological presuppositions of the
heuristics-and-biases approach to the psychology of rationality.

The controversy was in part ignited by implicit ambiguities within the heuristics-and-biases program itself. While formal theories of rationality had provided the model of good reasoning during the zenith of the Cold War and had often been taken to be both descriptively and prescriptively valid, the heuristics-and-biases approach undermined their descriptive adequacy, though not their prescriptive legitimacy. Rationality—understood as a system of rules—was still conceived of as purely formal. Psychologists expended little consideration about when it was, and when it was not, appropriate to apply the rules to the tasks given to subjects, and little curiosity about whether subjects had epistemic access to the rules when they tried to solve the problems. And rarely did psychologists reflect on the authority of the rules used in the test. Insofar as they ignored these issues, they continued to hope that reason could and should be exercised in more or less automatic ways.

The earliest serious protests came from outside of psychology, from the discipline that traditionally views itself as, in Kant's metaphor, the lawgiver of reason. The Oxford philosopher Laurence Jonathan Cohen (1923–2006) was the first to express severe dissatisfaction with the irrationality message and with many of the conceptual and methodological assumptions of the program. He began his career as a political philosopher and in 1954 published *The Principles of World Citizenship*. In the 1970s he studied inductive inference, including its application to law and medicine. His first attacks on the heuristics-and-biases program came in 1977, and in 1979 and 1980 he engaged in the first of several debates with Kahneman and Tversky on the issue of sample sizes in human predictions, published in the influential journal *Cognition*.[50] A year later, Cohen published a "target article" in *Behavioral and Brain Sciences*, flanked by twenty-nine invited responses by philosophers and psychologists, many of them leading figures in their fields.[51] Wason, Kahneman, and Tversky took the opportunity to respond as well.

Cohen's main point was no less challenging than the irrationality message itself: human irrationality could *not* be demonstrated experimentally. He employed the linguist Noam Chomsky's distinction between performance and competence: ordinary human speakers may commit linguistic mistakes and may have difficulties in learning languages; but the clear evidence points in favor of the assumption that they will be able to learn in principle any natural language if placed in the right environment. Humans have an innate language competence. The same, Cohen contended, held for rationality. Reasoning tests like the WST and the Linda problem showed merely that normal adults occasionally made mistakes, not that they were systematically and irremediably doomed to irrationality. On the contrary, the very norms of reasoning were expressive of ordinary human reasoning competence. To identify these norms, a subtle procedure of a "reflective equilibrium" would have to be applied:[52] an iterated application of balancing basic intuitions about particular inferences with general principles, until one achieved a comprehensive normative theory.[53]

Cohen's intervention marked a turning point in discussions of the psychology of rationality. Alongside Cohen, albeit with different arguments, several philosophers and psychologists began to call foul: there must be something wrong, they claimed, if science shows that human beings are, by and large, irrational. If we're so dumb, how come we made it to the moon?[54] Why did the species not become extinct long ago? This debate attracted much attention and was variously described as "fundamental,"[55] and eventually as the "rationality wars."[56] It remains unresolved. In its wake, psychological research on rationality fragmented.[57]

Already in the 1970s and '80s, three kinds of objections were advanced against the inference to widespread and ineradicable human irrationality. First, the followers of this program overemphasized negative data at the expense of existing positive data. Second, the allegedly ineradicable cognitive illusions could often be eliminated. Third, defenders of the heuristics-and-biases program assumed that the formal norms applied in the reasoning tests were unproblematic, whereas one could often show that they were inappropriately applied to problems or, alternatively, that some of the norms used in the tests were problematic themselves.

Objection 1: One-Sided Presentation of Empirical Results

For both logical and probabilistic or statistical reasoning tasks, there existed a vast literature that pointed in much more positive directions. Concerning logic, for instance, Rips and Marcus tested a number of simple syllogisms, both deductively valid and invalid ones.[58] All subjects correctly judged the *modus ponens* ($1. p \rightarrow q$; $2. p$, therefore q) as valid, and recognized that concluding not-q from its premises is invalid. They had difficulties with other syllogisms. But it was one-sided to claim that human beings are incapable of reasoning according to at least some of the rules of logic.

Similarly, with respect to probabilistic or statistical reasoning, a survey of studies by Peterson and Beach[59] supported the approximate empirical adequacy of the idea of "man as an intuitive statistician," a metaphor coined by Egon Brunswik (1903–1955). Among other issues, they reported on empirical studies of judgments of proportions, means and variances, statistical inferences or the drawing of sample sizes. One may wonder whether it is sensible to ask laypersons questions that require some statistical training, but subjects still did not do so badly. Peterson and Beach claimed that "the normative model provides a good first approximation for a psychological theory of inference."[60] They did not overlook discrepancies between normative theory and actual behavior. Authors in the heuristics-and-biases tradition, however, charged Peterson and Beach with being overly optimistic.[61] A later survey by Christensen-Szalanski and Beach[62] counted the studies sending either a more optimistic or a more pessimistic message about the responses of subjects in various reasoning tests. Of 84 studies, 37 reported good and 47 bad performances. The latter reports were cited much more often—about 27.8 times in contrast to 4.7 times the number of citations of studies of good performance within the time period studied. Psychologist Lola L. Lopes, another early critic of the heuristics-and-biases program,[63] claimed that it was in particular the appearance of an article by Tversky and Kahneman[64] in *Science* that shifted the balance of citations.[65]

Objection 2: Apparent Errors or Fallacies Can Be Eliminated

A second important objection was that the alleged fallacies *are* avoidable. For example, certain (not all) *contents* helped to improve reasoning. One such "content effect" was noted first by Griggs and Cox.[66] It concerned *deontic* rather than *descriptive* conditionals, that is, "if-then" sentences that express duties, permissions or prohibitions rather than factual assertions. Consider rules such as the following:

"If you take the benefit, then you have to pay the cost."
"If someone is under 18 years old, (s)he is required to drink Coke
(a nonalcoholic beverage)."

Once again, experimental subjects were presented with the question of what cards have to be turned over to check the observance of the rules (figure 6.4).

The main result here was that cheaters (i.e., those who take the benefit but do not pay the costs, or underage drinkers) were detected by more than

---

Four Cards:

$\boxed{\text{benefit}}$ $\boxed{\text{no benefit}}$ $\boxed{\text{cost paid}}$ $\boxed{\text{cost not paid}}$

Statement: *If you take the benefit, then you have to pay the cost.*

Question: What card(s) has (have) to be turned over to see if this statement has been violated?

Figure 6.4. A deontic version of the WST.

75 percent of experimental subjects.[67] This appeared to fit much better with what the material implication requires, and so it was often thought that certain contents facilitate reasoning in conformity with logic.

Such multifarious data cried out for explanation. This perhaps helps to explain, in part at least, why quite new conceptions of rationality came into existence.

Objection 3: Difficulties with Norms

A third objection concerned the rules of reasoning accepted uncritically by defenders of the heuristics-and-biases program. This point is especially important for this book, given that it directly concerns the very rules one assumes are essential for rationality. The objection came in two different forms.

Sometimes it was argued that a rule was *misapplied* to problems where one could show that subjects may very well have been following another rule, and reasonably so. Cohen made this point with respect to logical examples, judgments of randomness, and the famous "gambler's fallacy" (the tendency to think that, after a long series of heads the probability of tails must be more than 0.5).[68]

The second version of this objection questioned the unreflective acceptance of certain algorithmic rules of logic, probability theory, and decision theory by adherents of the heuristics-and-biases program. History indeed shows that there is often more than just one theory or model within these formal systems, and that disputes exist which at least occasionally are deep and serious.[69]

Cohen argued that there was nothing sacrosanct about so-called standard theories of logic and probability. He attacked Bayesian theories of probabilistic inference, claiming that a "Baconian" theory—which emphasizes qualitative rather than quantitative aspects of evidence, presses more for variation rather than repetition of experiments, and focuses on the elimination of hypotheses instead of calculating the values of evidence for them—would provide a better alternative in, say, medical diagnosis.[70] Birnbaum also argued that to evaluate the credibility of witnesses' judgments in courts, mere invocation of base rates was insufficient and should be supplanted by proper perceptual theory.[71]

6.7. The Fragmentation of Rationality

While Kahneman and Tversky's work had emphasized the gulf between the normative and the descriptive, on the normative side they still adhered to what we have characterized as the ideal type of Cold War rationality: good reasoning has to follow formal algorithms that optimize results and can be applied mechanically. Under the pressure of the aforementioned (and other) critiques, however, notions of rationality began to fragment in the 1980s. Aspects of the ideal type were torn apart, and new conceptions introduced. To describe the variants in detail would be beyond the scope of this book, and the debates within psychology and philosophy are in any case still ongoing. However, three alternatives, here presented only in schematic outline, give some sense of the multiple directions taken in deliberations about how rationality came to be understood since then.[72]

First, some psychologists[73] claimed that humans have two irreducibly different systems of reasoning. Different versions of this approach proliferated, since the idea that there are two different reasoning systems was connected with various claims about properties of these systems: whether "system 1" and "system 2" were fast or slow, tacit or explicit, heuristic-based or rooted in formal rules, holistic or analytic, modular or central, and other distinctions as well.[74] Sometimes (though by no means always), it was also claimed that the first system is one for choosing the right means to attain desired ends, and the other for following norms of logic, probability theory, and other formal standards, thus drawing a sharp wedge between the instrumental and the formal aspects of rationality.

Second, several researchers further strengthened the idea that rationality is "bounded," usually with reference to the ideas of Herbert Simon,[75] as seen in chapter 2. In some versions, this simply meant recognizing that our cognitive capacities and resources are limited: neither computational power nor memory nor available time is infinite. Kahneman, Gilovich, and others claimed to be following Simon's path.[76] Berlin psychologist Gerd Gigerenzer,[77] however, questioned how faithful these interpretations were to Simon's conception. Simon had illustrated his view with the metaphor of a pair of scissors: "Human rational behaviour is shaped by a scissors whose two blades are the structure of task environments and the computational capabilities of the actor."[78] For Simon, moreover, bounded rationality could under certain conditions be normatively as well as descriptively valid. "Satisficing" was not merely what agents *did* when they could not optimize according to expected utility theory; rather, it was what they *should* do, namely rest satisfied once a certain aspiration level had been reached, full stop (see chapter 2). Gigerenzer's own conception went in the same direction. To figure out the fitness between norms and contents and contexts of their use is a matter of empirical research. Evidence for this approach consisted in the construction of so-called fast-and-frugal heuristics, simple reasoning strategies that work in specifiable domains and only in these. In contrast to Kahneman and Tversky, Gigerenzer and his collaborators viewed heuristics not as typically defective. Despite their simplicity, many of them often did as well as, or even better than, complex formal rules of judgment and decision making.[79] However, applying heuristics *successfully* involves a certain kind of mindfulness: one has to be able to select consciously which heuristic is the most reasonable one for a given task and context.

Third, others argued for a radically evolutionary conception of rationality. A departure point of their empirical research was the task of explaining the results in the deontic versions of the WST and similar reasoning tests. Drawing on the work of Robert Trivers (see chapter 5) and others[80] concerning reciprocal altruism, psychologist Leda Cosmides and anthropologist John Tooby argued as follows. At some point during the evolution of *Homo sapiens*, social cooperation became highly advantageous. This behavior, however, can only be an evolutionarily stable strategy if humans possess the cognitive ability for spotting cheaters. Those who lacked this ability would be open to exploitation and be selected against. Thus, natural selection led to the development of a "cheater detection module": people are quite able to detect who violates certain social rules, as in the drinking or cost-benefit rules mentioned above. At the same time, there has been no comparable selection pressure for understanding the logic of material implication or other formal rules.[81]

All three of these research programs are, of course, contested, as are the abovementioned objections to the heuristics-and-biases approach. A many-sided debate about the nature of rationality and irrationality has been raging at the interfaces of psychology and philosophy, economics, and biology since the mid-1980s. Yet if we turn to the applications of the psychology of rationality to policy analysis, hints of dissent within the ranks are very rare. The monolithic definitions of rationality and irrationality inherent to the heuristics-and-biases school of thought still hold sway.

Kanwisher, for example, defended her views about the weaknesses of American security policy by referring to Kahneman and Tversky's studies on conjunction errors and overconfidence.[82] She did so at a time when several crucial objections to their approach had already been voiced fully and clearly.[83] But these critics were not mentioned in her work. Nor did she reflect on the point that the professional forecasters and planners might view the combined possibility of "a Russian invasion of Poland, and a complete suspension of diplomatic relations between the USA and the Soviet Union, sometime in 1983" as more likely because of possible causal dependencies—after all, how would Reagan have reacted to such an invasion? Kahneman and Tversky themselves had speculated that an invasion might be viewed as a cause leading to the breakdown of diplomatic relations.[84] They did not make much of this point either. The conjunction rule requires, however, that the two conjuncts describe *independent* events. Moreover, is it adequate in the first place to present forecasters with simplified alternatives when events in a domain are frequently more complex and conditional upon each other?[85]

Nor did McDermott, in her analysis of Reagan's mindset, question her own implicit assumption: the idea that there was some normative rule of probability that would permit exact predictions about the Soviets' future behavior.[86] That political relations might just be too complex to be calculable this way was never considered. Not surprisingly, McDermott did not make explicit which formal rule of rationality Reagan should have used instead of his set of firmly held beliefs and values. Like Kanwisher, although writing over a decade later, McDermott ignored the heated debate over the explanatory power of heuristics like availability or representativeness, which had already been challenged as vacuous.

To be fair, not all analysts are equally unreflective or naïve about the potentials of heuristics and biases and its concomitant normative commitment to formal theories of reasoning. Among the (partial) exceptions are the works of Peter Suedfeld and, more influentially, Philip E. Tetlock, both psychologists by training but with a strong interest in political topics. Suedfeld has claimed that international relations do not straightforwardly allow for a mechanical application of formal theories of logic or of cost-benefit calculations—for instance, moral or social values may outweigh them, both descriptively and normatively. Therefore, we "should stop calling preferences or tendencies 'errors.'"[87] He also argued for a better distinction between standards of formal theories and standards of success, thus separating different aspects of rationality. Finally, he maintained that good decision makers need "metadecisions (i.e., deciding what strategy to adopt and how much time and effort to expend on particular decisions)."[88]

Tetlock, in turn, is known for his comprehensive study of political expertise, which began in the late 1980s. In a field study conducted over twenty years with several hundred government officials, university professors, and journalists he raised grave doubt about their ability to predict political developments.[89] Tetlock distanced himself from the heuristics-and-biases program in several respects. For instance, with respect to political forecasting, he distinguished between three views:[90] "skeptics," "complexifiers," and "fundamentalists." Skeptics think successful predictions are either the result of sheer luck or the careful application of probability theory (so this group is, by and large, identical with defenders of heuristics and biases). Complexifiers think that key to overcoming biases is "preemptive self-criticism," and that the ability to make good predictions is rooted in knowing all the fine-grained details of politics. Fundamentalists, finally, claim that looking for the major "underlying geopolitical, economic, and technological forces" is the right thing to do, and that one can ignore short-term "distractor variables that dominate the headlines."[91] Now, Tetlock claimed that one should not adopt one approach at the expense of the others but instead apply a kind of metacognition for selecting which of the three views is appropriate in a given context: "the ability of observers to shift styles of thinking in response to situational demands."[92] Hence, he did not accept the heuristics-and-biases program tout court. Still, he emphasized that the major result of his field study vindicated the skeptic: political forecasters do only a little better than chance. He explicitly described his results as "heartening to those in the judgment and decision-making community who believe that demonstrations of systematic errors and biases are not just byproducts of laboratory trickery performed on unmotivated undergraduate conscripts."[93] But the fundamental objections presented above (in section 6.6) remained ignored.

It is not obvious that applications of the psychology of reasoning to the political domain will be improved by considering all objections to the heuristics-and-biases program, or by exploiting one or another of the alternative conceptions of rationality. But that is as it should be. To understand what combination of events was probable in the early 1980s requires extremely hard thinking. That the Polish crisis could, rather than revive the Cold War, be a step toward overcoming the remnants of this era was difficult to see for contemporaries. In 1970, the Russian writer and dissident Andrei Amalrik had predicted the downfall of the Soviet Union by 1984.[94] By the mid-1980s he could surely still be seen as blatantly mistaken. By 1991, such a prediction might have appeared as almost prophetic. Of course, one could also see it as based on sheer luck: "the abrupt end of the Cold War astonished almost everyone."[95] No rule, whether of logic or probability theory or rational choice theory, mechanically applied, is likely to be able to handle the gamut of political choices. In a curious way, heuristics and biases, like the rules of game theory or rational choice theory, held out the promise of being applicable everywhere and always, and were applied by researchers in an often equally unreflective way.

What is more, researchers following this approach did not consider whether what looks from one point of view as irrational behavior might appear rational from another perspective. The dichotomy between rationality and irrationality has therefore come to resemble what St. Augustine once said about time: we seem to know what it is when not being asked to say what it is, but once we are asked it becomes evident that we do not really know. In the wake of the disputes over the heuristics-and-biases program and its critics, it is no longer clear what counts as rational or irrational, because the options (rules and standards) have multiplied and diverged. We do not yet know with what to replace the ideal of Cold War rationality.

Cold War Rationality after the Cold War

The Cold War is long over, and the debate over Cold War rationality ended with it—more or less. Our story has not been an epic of rise and fall, life and death, but rather one of condensation under intense pressure: people, ideas, disciplines, methods, and institutions that dwelled apart both before and after the Cold War briefly came together. This book has been about a set of debates during the Cold War that attempted both to frame and develop the problem of rationality under the most pressing, most anxious, and yet perhaps most propitious circumstances. We began with a picture of an ideal type of rationality, one that focused on formal rules, algorithms, optimization, and mechanization. As we saw in the first two chapters (which brought the story to the early 1950s), this was only the starting point for a vigorous debate about what rationality might be and how it could be operationalized, made useful in a very tense international climate. Over the following decades, American intellectuals from across the human sciences went at that pared-down ideal type, pushing and stretching it in order to render it a better fit with the empirical world. The next chapters traced precisely those struggles to make a usable rationality, one that actually fit the behavior of human beings (or dung flies), until debates within psychology and philosophy in the late 1970s and early '80s vitiated the debate from within. There are still arguments over rationality, of course, but those features that made a discourse out of separate debates, which framed the rules of the game in a manner that we have labeled Cold War rationality, are absent. To be sure, there are schools of rational choice theorists in political science; operations research and management science are taught in business schools and deployed in government and industry; game theory, both theoretical and experimental, is a vast subfield within economics; Bayesian decision theory has been applied to everything from weather prediction to the assessment of scientific evidence; and the "strange situation" is everywhere, from social psychology labs to focus groups. But the debate that gave Cold War rationality its energy is over.

And what an energy it was! The assumptions might seem bizarre, and the hubris almost certainly off-putting, but there was definitely nothing stale about the arguments that raged across the human sciences in this period, spanning "the whole set of concepts associated with subjective probability, maximizing behavior, solution concepts for cooperative and non-cooperative games, and so on"[1]—which certainly doesn't sound that glamorous anymore. But project yourself back to 1966, when that statement was made. You are sitting in a room with some really smart people. Not just smart; *the smartest ones around*. They hailed from every field in the human sciences, they were at the peak of their intellectual powers, and they came to the same conferences, seminars, and meeting rooms not to defend some square cubits of disciplinary space but to engage in the reformulation of what rationality meant. The ideal type of minimalist rationality was the starting point, not the final result, and they rolled up their sleeves and talked about anything and everything, to anyone who would listen.

People certainly did listen. And why not? It was riveting. Herman Kahn outlined his thermonuclear scenarios and then Thomas Schelling undercut them, with Anatol Rapoport clamoring for even more assumptions to come under the knife. A very specific, very contingent set of circumstances coalesced during the height of the Cold War to make these debates over rationality not only a fascinating conversation, but the most important conversation across the gamut of the human sciences in the United States. First, there were the enormous geopolitical stakes. Rationality, all were convinced, was the only means to prevent the insanity of global nuclear war and the extinction of man—now *there* was a billet to focus the mind. And then came the money, tsunamis of funding (much of it military) that washed across the various disciplines to foster this kind of interdisciplinary conversation about exactly this question. As a direct result of the scale of the task and the lavish opportunities it offered, the elite of the disciplines scrambled into the fray. They were clever and most were young, relishing the possibility of remaking entire disciplines before they had calcified into stodgy old dogmas. Many of them also had real-world experience of how to accomplish serious tasks—either in World War II or as advisors to postwar reconstruction programs—and thus were equipped with a practical as well as a theoretical eye. The right people at the right stage of their careers with the right problem and enough funds to give it a go—is it any wonder they were captivated?

Yet today much of the glitter has worn off. The mainstays of the debate over Cold War rationality have fragmented into academic specialties, more or less respectable and more or less robust—and quite a bit less connected to one another. From the beginning, all the participants in these debates had focused on the minimalism of the ideal type. For proponents of a formal, algorithmic rationality, the notoriously mean and lean *Homo economicus* was a model of the stripped-down approach to human rationality in general, not just in the market and the stock exchange. Human scientists should make as few assumptions as possible, those few should be self-evidently derived from self-interest, and they should yield explanations and predictions for a wide range of actual human behavior. In pleasingly paradoxical fashion, such a thin theory of rationality would disprove the economists' own axiom that "you can't get something for nothing": a paltry set of rules would yield a bumper crop of results. As Schelling put it in the midst of a fierce debate over the meaning of rationality at the 1964 Berkeley conference on conflict strategies, the economists' theory of rationality provided "free of charge, a lot of vicarious, empirical behavior. Where there are no strong personality and emotional determinants of behavior, people on the whole do the 'right' [i.e., rational] thing."[2] Although acutely aware of the limitations of skeletal rationality, stripped of all "strong personality and emotional determinants"—limitations pointed out by numerous opponents whenever rationality was debated—Schelling and some of his colleagues nonetheless hoped that it could be fruitfully extended to the human sciences in general.

That aspiration is what has changed. Animating the debate over Cold War rationality was an engagement both with the algorithmic formal presentation of rationality *and* the criticisms voiced against it: both John Nash and Charles Osgood, both Herman Kahn and Herbert Simon. The debate has ceased to be interdisciplinary, has been channeled into well-worn paths marked by agreed-on premises. No one begged the question anymore, as had been routine during the Cold War. The critics and antagonists of Cold War rationality, for their part, moved their arguments into their own disciplinary tracks: theories that promoted "local context" and "thick description" now assumed their places within these fields.[3] As the Cold War tensions eased and the lavish fountains of money for the human sciences dried up, Cold War rationality ceased to be a single battlefield of contestation. The kinds of academics and intellectuals who had been part of the original disputes discussed in this book, who hailed from a wide variety of disciplines, now talked mostly to their peers in their own discrete fields, no longer to one another nor to the high echelons in the Pentagon on a quest to remake the world. There were intellectual consequences, too. It was no longer obvious, for example, what game theory and Bayesian decision theory had to do with one another; they could even be seen as competing accounts of the same phenomena. In retrospect, the expansive "so on" of the 1964 definition of rationality read as an invitation to miscellany.

Cold War rationality has nonetheless left a deep if shadowy imprint on the post–Cold War intellectual landscape. Even if what was once a single if fractious discussion is now broken up and dispersed all over the academic map, the slimmer version of Cold War rationality that ignited the debate still defines by opposition an equally distinctive form of irrationality, studied by the discipline of psychology. As the foregoing chapters document, Cold War rationality never lacked sharp critics, many of the sharpest coming from those most committed to the idea that rationality, new and improved, might rescue the world from nuclear self-destruction. Time and time again they tried to put some psychological flesh on the bare-bones models with the aim of achieving a more rational rationality, equal to the demands of an all-too-real world of mutual assured destruction. Sometimes, as we saw in chapters 3 and 5, this took the form of expanding the meaning of the rational to include the revision of prejudices in the face of cognitive dissonance or mutual trust or moral commitment. Sometimes, as we saw in chapter 4, psychology entered discussions of rationality by introducing carefully framed and controlled social interactions. And sometimes, as we saw in chapters 5 and 6, more psychology came by way of more empiricism, for example, in experimental game theory or the results of research on heuristics and biases, both of which seemed to show that subjects systematically deviated from rational norms.

More psychology did not necessarily mean more irrationality: some human scientists insisted that rationality be redefined to take account of psychology (including computational constraints), a position forcefully argued by some game theorists, economists, political scientists, and philosophers, as well as psychologists, as we have seen in chapters 2, 3, 4, and 5. But the success of the research program known as "heuristics and biases"[4] in psychology narrowed the definition of rationality to a still barer version (such as a few rules of standard logic and Bayes's theorem) of bare-bones rationality (as shown in chapter 6) and thereby froze the connection between psychology and the investigation of irrationality. And not just any kind of irrationality: the peculiar brand of irrationality investigated from the 1970s onward had been defined as the photographic negative of the ideal type of Cold War rationality sketched in our introduction—violations of preference consistency, Bayesian probabilities, maximizing behavior, game theoretical strategies, and so on. It is no coincidence that it was the discipline of experimental psychology that internalized the precepts of Cold War rationality most deeply in its own methods, as we saw in chapter 4. The irrationality of the experimental subjects was simply the inverse of the psychologists' own rational methods—many of which were forged in the Cold War crucible.[5]

In contrast, many Cold War rationalists had confidently assumed that most people were rational, in just the sense that their own pet methods and theories defined rationality. John Q. Public (and Ivan Ivanovich Ivanov) maximized (or at least satisficed), revised their Bayesian priors in light of evidence, and kept their preferences tidily transitive. They not only played by the rules; they thought by them, insofar as any thinking was required in executing algorithms. As we saw in chapter 1, this rigidly rule-bound vision of rationality broke with older philosophical ideals of reason over the necessity and desirability of mindfulness. It is therefore at first glance surprising that this brand of rationality has become a specialty within the academic discipline of philosophy. The contributors to *The Oxford Handbook of Rationality* (2004), for example, are all philosophers; their articles include "Rationality and Game Theory," "Bayesianism," "Economic Rationality," "Rationality and Science," and "The Rationality of Being Guided by Rules," as well as those that examine classical texts in the history of philosophy (e.g., Hume and Kant) in the light of contemporary conceptions of rationality.[6] In the last fifty years, rationality has taken its place beside reason— just as, in the Enlightenment, reason took its place beside wisdom—as a subject matter (and perhaps also the method) of philosophy.

We began this epilogue with the crashingly banal assertion that the Cold War is over. Yet the nuclear weapons that made it a world-historical moment are still there, still packed into the noses of intercontinental missiles pointed at Moscow and Washington DC. So why has the wide-ranging, free-wheeling, high-stakes debate over Cold War rationality collapsed? In conferences and seminar rooms in university departments across the human sciences there are still no doubt vigorous discussions over this or that aspect of game theory or rational choice theory, economic optimization versus satisficing, the psychology of judgment and decision making, or even nuclear strategy. But no matter how intense, these are mostly specialist discussions conducted within disciplines, not cutting across them. The methods, examples, applications, and citation patterns all bear the stamp of various disciplinary standards and preoccupations. Still more striking than disciplinary specialization is the lack of interest in expanding and enriching formal rationality, as opposed to rejecting it outright. Neither rational choice theory nor game theory nor economic optimization models lack for critics who condemn them and all their works. But this is a far cry from the Cold War rationality debate, whose participants took for granted a shared starting point for formulating a new, improved rationality equal to the challenges of preserving the world from nuclear immolation. It is easy to find fragments of the debate in this or that subdiscipline; it is all but impossible to find anything like the debate itself—its depth, its breadth, its exhilarating and terrifying sense of urgency.

We have attempted an intellectual history that goes beyond the fragments, fascinating though each may be as embedded in its past and present disciplinary matrix. The ideas, methods, and practices that constitute the fragments have genealogies; they can and should be traced. But the debate that once fit them all together has no such continuity. Neither does its neon megawattage, which attracted the keenest minds like moths. The contemporary equivalents of *Life* and *Business Week* no longer feature admiring portraits of "action intellectuals" and "Pentagon planners," though these types are still alive and well. This is the kind of historicity that we have tried to capture, the constellation of minds and matters that suddenly link up and light up ideas that exist separately both before and after the particular historical moment. The history of reason is punctuated by such electrified moments, which can be only partially grasped by the history of ideas or even a more contextualized disciplinary history. The history of the debate over Cold War rationality was one such episode, and perhaps the highest voltage one to date.

For approximately forty years, the history of rationality was forged in flight plans for feeding West Berlin, tense deliberations of the Cuban Missile Crisis, war games played between generals and strategists, "situations" devised by social scientists in Micronesia and at Harvard, and in other contexts seldom frequented by academic philosophers. This was rationality stripped for action in an age of high drama: for all anyone knew, the world might end tomorrow with a very big bang. Swift, calculable moves and countermoves would safeguard a dangerously precarious balance—or so was the hope. To many of the best and the brightest, mindful reason, subtle but slow, seemed a luxury that nations brandishing nuclear weapons at each other could not afford. If in retrospect, from a comfortable distance and safe in the knowledge that the Cold War did not in fact erupt into the hottest war in human history, the drama looks more like melodrama—at times even rendered as farce—that is simply proof of how extraordinary, and extraordinarily strange, those times were.

NOTES

INTRODUCTION

1. Peter Bryant [Peter George], *Red Alert* (Rockville, MD: Black Mask, [1958] 2008), 160.

2. On the impact of Cold War novels and films, see Margot A. Henriksen, *Dr. Strangelove's America: Society and Culture in the Atomic Age* (Berkeley: University of California Press, 1997).

3. Thomas C. Schelling, "Meteors, Mischief, and War," *Bulletin of the Atomic Scientists* 16 (1960): 292 (original emphasis).

4. The portmanteau terms used to describe the nonnatural sciences varied, often in interesting and controversial ways: what in the early nineteenth century had been known in French as the *sciences morales*, rendered into English by John Stuart Mill as the "moral sciences" and then translated into German as the *Geisteswissenschaften*, became in English the "social sciences" by the mid-nineteenth century, further differentiated into the "social" and "behavioral" sciences (the latter referring chiefly to the use of psychology to build bridges across sociology, social anthropology, history and economics) in the 1950s (Wilhelm Dilthey, *Einleitung in die Geisteswissenschaften* [1883] in Wilhelm Dilthey, *Gesammelte Schriften*, ed. Bernhard Groethuysen, v. 1 [Stuttgart/Göttingen: B.G. Teubner/Vandenhoeck & Ruprecht, 1990], 5; Ernst Rothacker, *Logik und Systematik der Geisteswissenschaften* [Bonn: H. Bouvier u. Co. Verlag, 1947], 4–16). The German translation of Mill's *A System of Logic, ratiocinative and inductive* (1843) appeared in 1849. See also Theodore M. Porter, "Genres and Objects of Social Inquiry, from the Enlightenment to 1890," in *The Cambridge History of the Modern Social Sciences*, ed. Theodore M. Porter and Dorothy Ross, 13–39 (Cambridge: Cambridge University Press, 2003); and Dorothy Ross, "Changing Contours of the Social Science Disciplines," in *The Cambridge History of the Modern Social Sciences*, 205–37. Because these classification systems were continually debated and redrawn (was, e.g., psychology a natural, social, or behavioral science?), in this book we will use the more inclusive and neutral "human sciences" to refer to all disciplines that systematically study human phenomena (which can encompass some parts of both biology and philosophy at its boundaries, as we will see).

5. Mary S. Morgan, "Economic Man as Model: Ideal Types, Idealization and Caricatures," *Journal of the History of Economic Thought* 28 (2006): 1–27.

6. There were also plenty of alternative discussions about unification of the human sciences and reformulation of the public sphere in their image that had nothing to do with these Cold War debates, such as the virtual obsession with "creativity" and "interdisciplinarity." See Jamie Cohen-Cole, "The Creative American: Cold War Salons, Social Science, and the Cure for Modern Society," *Isis* 100, no. 2 (2009): 219–62.

7. Thomas C. Schelling, "Uncertainty, Brinkmanship, and the Game of Chicken," in *Strategic Interaction and Conflict*, ed. Kathleen Archibald, 74–87 (Berkeley: International Security Program, Institute of International Studies, University of California at Berkeley, 1966), 81.

8. See the entries "Rationalität, Rationalisierung I/II/III," in *Historisches Wörterbuch der Philosophie*, vol. 3, ed. Joachim Ritter, Karlfried Gründer, and Gottfried Gabriel, 42–66 (Basel: Schwabe, 1971–2007); and "Vernunft/Verstand" and "Vernunft, instrumentelle," in Ritter, Gründer, and Gabriel, *Historisches Wörterbuch der Philosophie*, vol. 11, 748–866 and 866–67.

9. John Rawls, "Kantian Constructivism in Moral Theory," *Journal of Philosophy* 77 (1980): 515–72. Rawls was influenced by W. M. Sibley, "The Rational versus the Reasonable," *Philosophical Review* 62 (1953): 554–60; cf. Rawls, *Political Liberalism*, rev. ed. (New York: Columbia University Press, 1996), 47–88, esp. 50–54. Rawls himself noted that the distinction marked a substantive revision of his earlier views: in *A Theory of Justice* (Cambridge, MA: Harvard University Press, 1971), he had claimed that the theory of justice was part of rational choice theory. See also Hilary Putnam, *Reason, Truth and History* (Cambridge: Cambridge University Press, 1981), chapter 8. For the frequent positive adoptions of the term "rationality" by ethical anti-egoists and anti-instrumentalists see Henry Sidgwick, *The Methods of Ethics* (London: Macmillan, 1874), I.1.i; Thomas Nagel, *The Possibility of Altruism* (Princeton, NJ: Princeton University Press, 1970), 3; Derek Parfit, *Reasons and Persons* (Oxford: Oxford University Press, 1980), 3.

10. Thomas C. Schelling, "Discussion. First Session: The Concept of Rationality," in Archibald, *Strategic Interaction and Conflict*, 147.

11. On this point, see the thoughtful essay by Hunter Heyck, "Producing Reason," in *Cold War Social Science: Knowledge Production, Liberal Democracy, and Human Nature*, ed. Mark Solovey and Hamilton Cravens, 99–116 (New York: Palgrave, 2011).

12. On minimax theorems in the context of game theory, see Robert W. and Mary Ann Dimand, "The Early History of the Strategic Theory of Games from Waldgrave to Borel," in *Toward a History of Game Theory*, ed. E. Roy Weintraub, 15–28 (Durham, NC: Duke University Press, 1992). On utility theory, see George Stigler, "The Development of Utility Theory, Parts I and II," *Journal of Political Economy* 58 (1950): 307–27, 373–96; and Nicola Giocoli, *Modeling Rational Agents from Interwar Economics to Early Modern Game Theory* (Cheltenham, UK: Edward Elgar, 2003), chapter 2. On Bayes's theorem, see Stephen M. Stigler, "Thomas Bayes's Bayesian Inference," *Journal of the Royal Statistical Society* (A) 145 (1982): 250–58, and Stigler, *The History of Statistics: The Measurement of Uncertainty before 1900* (Cambridge, MA: Harvard University Press, 1986), chapter 3.

13. One should however be cautious about jumping to the conclusion that military funding was the prime mover behind these developments. For one thing, postwar military budgets both expanded and contracted under political pressures at different moments, nor were all military leaders enamored of the analytic approaches recommended by intellectuals with little or no battlefield experience. Moreover, the

interactions of the human sciences within academia and with the ambient culture created their own dynamic. See Hunter Heyck and David Kaiser, "Introduction: New Perspectives on Science and the Cold War," *Isis* 101 (2010): 362–66; and David Engerman, "Social Science in the Cold War," *Isis* 101 (2010): 393–400.

14. Theodore H. White, "The Action Intellectuals," with photographs by John Lonegard, *Life*, June 9, 1967, 3, 44, 64.

15. "Planners for the Pentagon," *Business Week*, July 13, 1963, 10.

16. The Institute of International Studies at the University of California at Berkeley was founded in 1955 to advance comparative (i.e., geopolitical) perspectives in the social sciences, especially as applied to contemporary problems. Its projects included the Comparative National Development Project and the Faculty Seminar on the Communist World.

17. The published proceedings of the conference included a transcript of the discussions as well as texts of the papers: "In editing the proceedings, an attempt has been made to retain the informality and spontaneity of the interchange, and the idiosyncrasies of presentation and of personality." Kathleen Archibald, introduction to Archibald, *Strategic Interaction and Conflict*, vi.

18. From Schelling's Nobel Prize autobiography, available at http://www.nobelprize .org/nobel_prizes/economics/laureates/2005/schelling-autobio.html (accessed August 19, 2011). See also Robert Dodge, *The Strategist: The Life and Times of Thomas Schelling* (Hollis, NH: Hollis Publishing, 2006).

19. Available on RAND's website, among other places, at: http://www.rand.org/about/ history/wohlstetter/P1472/P1472.html (accessed August 19, 2011). For more of Wohlstetter's writings, see the bibliography available on the above website and Robert Zarate and Henry Sokolski, eds., *Nuclear Heuristics: Selected Writings of Albert and Roberta Wohlstetter* (Carlisle, PA: Strategic Studies Institute, US Army War College, 2009).

20. Albert Wohlstetter, "Comments on Rapoport's Paper: The Non-Strategic and the Non-Existent," and Anatol Rapoport, "Rejoinder to Wohlstetter's Comments," in Archibald, *Strategic Interaction and Conflict*, 107–34.

21. Richard Bellman, *Eye of the Hurricane: An Autobiography* (Singapore: World Scientific, 1984), 136–37.

22. See, e.g., Alex Abella, *Soldiers of Reason: The Rand Corporation and the Rise of American Empire* (Orlando, FL: Harcourt, 2008).

23. "Editor's Note," *Life*, June 9, 1967, 3. For suggestive descriptions of the congenially argumentative climate cultivated by elite Cold War intellectual centers, see the description of RAND in Sharon Ghamari-Tabrizi, *The Worlds of Herman Kahn: The Intuitive Science of Thermonuclear War* (Cambridge, MA: Harvard University Press, 2005), 46–60, and the account of the early days of the Harvard Center for Cognitive Studies in Jamie Cohen-Cole, "Instituting the Science of Mind: Intellectual Economies and Disciplinary Exchange at Harvard's Center for Cognitive Studies," *British Journal for the History of Science* 40 (2007): 567–97.

24. Cultural pessimism about the unreasoning masses was strong among refugee intellectuals from Central Europe, such as Leo Strauss and Friedrich von Hayek, but did not seem to be especially salient among the mostly younger Cold War rationalists. Nor did they seem much swayed by fears about stupefied publics prophesized by cultural critics of the new media. See Richard Butsch, *The Citizen Audience: Crowds, Publics, and Individuals* (New York: Routledge, 2008).

25. Both quoted in "Planners for the Pentagon," 10, 11.

26. White, "Action Intellectuals," 44.

27. John von Neumann and Oskar Morgenstern, *Theory of Games and Economic Behavior* (Princeton, NJ: Princeton University Press, 1944).

28. Oskar Morgenstern, "Some Thoughts on Maxims of Behavior in a Dynamic Environment," (letter to *Forum for Contemporary History*, 1974), Oskar Morgenstern Papers, Box 32, "Some Thoughts on the Maxims of Behavior in a Dynamic Environment 1975–1976" folder, Duke University Special Collections, Durham, NC.

29. George Kennan to Oskar Morgenstern, August 3, 1974, Oskar Morgenstern Papers, Box 32, "Some Thoughts on the Maxims of Behavior in a Dynamic Environment 1975–1976" folder, Duke University Special Collections, Durham, NC.

30. Anatol Rapoport, "Rejoinder to Wohlstetter's Comments," in Archibald, *Strategic Interaction and Conflict*, 127.

31. Robert Leonard, *Von Neumann, Morgenstern, and the Creation of Game Theory* (Cambridge: Cambridge University Press, 2010), 284–86.

32. E. Vilkas, ed., *Uspekhi teorii igr: Trudy II Vsesoiuznoi konferentsii po teorii igr (Vil'nius 1971)* (Vilnius: Mintis, 1973), 5.

33. Quoted in N. N. Vorob'ev, "Sovremennoe sostoianie teorii igr," in *Teoriia igr*, ed. N. N. Vorob'ev, et al., 5–57 (Erevan: Izd. AN Armianskoi SSR, 1973), 5.

34. N. N. Vorob'ev, "Prilozheniia teorii igr (Metodologicheskii ocherk)," in Vilkas, *Uspekhi teorii igr*, 250.

35. N. N. Vorob'ev, "Nauchnye itogi konferentsii," in Vilkas, *Uspekhi teorii igr*, 8–9.

36. *Against the Philosophizing Henchmen of American and English Imperialism* (1951) included a paper by Mikhail Iaroshevskii that attacked cybernetics as idealism under the rubric of "semantic idealism," a holdover from the linguistics wars of the previous year. See Slava Gerovitch, *From Newspeak to Cyberspeak: A History of Soviet Cybernetics* (Cambridge, MA: MIT Press, 2002), 120–21.

37. Atsushi Akera, *Calculating a Natural World: Scientists, Engineers, and Computers during the Rise of U.S. Cold War Research* (Cambridge, MA: MIT Press, 2007).

38. Even game theory, parochial mathematical discipline though it was, was folded into cybernetics in the Soviet context: Vorob'ev, "Sovremennoe sostoianie teorii igr," 50–52.

39. Gerovitch, *From Newspeak to Cyberspeak*, 200.

40. The literature in all instances is vast, and the following are representative of some of the best work in these areas. For economics, see Philip Mirowski, *Machine Dreams: Economics Becomes a Cyborg Science* (Cambridge: Cambridge University Press, 2002). For cybernetics, see Peter Galison, "The Ontology of the Enemy: Norbert Wiener and the Cybernetic Vision," *Critical Inquiry* 21 (1994): 228–66; Lars Bluma, *Norbert Wiener und die Entstehung der Kybernetik im zweiten Weltkrieg* (Münster: Lit, 2005); and Andrew Pickering, *The Cybernetic Brain: Sketches of Another Future* (Chicago: University of Chicago Press, 2010). For artificial intelligence, see H. M. Collins, *Artificial Experts: Social Knowledge and Intelligent Machines* (Cambridge, MA: MIT Press, 1990). For military strategy, see Ghamari-Tabrizi, *The Worlds of Herman Kahn*; and Fred Kaplan, *The Wizards of Armageddon* (New York: Simon and Schuster, 1983). For game theory, see Mary S. Morgan, "The Curious Case of the Prisoner's Dilemma: Model Situation? Exemplary Narrative?" in *Science without Laws: Model Systems, Cases, Exemplary Narratives*, ed. Angela N. H. Creager, Elizabeth Lunbeck, and M. Norton Wise, 157–85 (Durham, NC: Duke University Press, 2007); and Leonard, *Von Neumann, Morgenstern, and the Creation of Game Theory*. For computers, see especially Paul Edwards, *The Closed World: Computers and the Politics of Discourse in Cold War America*

(Cambridge, MA: MIT Press, 1996). On cognitive science, see Margaret A. Boden, *Mind as Machine: A History of Cognitive Science*, 2 vols. (New York: Oxford University Press, 2006) and Hunter Crowther-Heyck, *Herbert A. Simon: The Bounds of Reason in Modern America* (Baltimore, MD: Johns Hopkins University Press, 2005). And for philosophy of science, see George Reisch, *How the Cold War Transformed Philosophy of Science: To the Icy Slopes of Logic* (Cambridge: Cambridge University Press, 2005). Further references are to be found in the following chapters.

41. Sonia M. Amadae, *Rationalizing Capitalist Democracy: The Cold War Origins of Rational Choice Liberalism* (Chicago: University of Chicago Press, 2003).

42. On funding, see Stuart W. Leslie, *The Cold War and American Science: The Military-Industrial-Academic Complex at MIT and Stanford* (New York: Columbia University Press, 1993); Christopher Simpson, ed., *Universities and Empire: Money and Politics in the Social Sciences in the Cold War* (New York: New Press, 1998), but also Engerman, "Social Science in the Cold War." For key figures, in addition to the studies cited in note 40 above, see Steve J. Heims, *The Cybernetics Group* (Cambridge, MA: MIT Press, 1991); Dodge, *The Strategist*; Flo Conway and Jim Siegelman, *Dark Hero of the Information Age: In Search of Norbert Wiener, the Father of Cybernetics* (New York: Basic Books, 2005); Crowther-Heyck, *Herbert A. Simon*; Giorgio Israel, *The World as Mathematical Game: John von Neumann and Twentieth-Century Science* (Basel: Birkhäuser, 2009); Peter Byrne, *The Many Worlds of Hugh Everett III: Multiple Universes, Mutual Assured Destruction, and the Meltdown of a Nuclear Family* (Oxford: Oxford University Press, 2010); and Leonard, *Von Neumann, Morgenstern, and the Creation of Game Theory*.

43. The historiography is too vast to cite in detail, but for an overview, tracking both the foreign conflicts and the reverberations within the United States, see Melvyn P. Leffler and Odd Arne Westad, *The Cambridge History of the Cold War*, 3 vols. (Cambridge: Cambridge University Press, 2010), and the many references cited therein.

CHAPTER ONE

1. The children were familiar with idea of a reverse auction from the book *Cheaper by the Dozen* (New York: T.Y. Crowell, 1948), a memoir by Ernestine Gilbreth Carey and Frank B. Gilbreth Jr. about growing up in the large family of efficiency experts Frank and Lillian Gilbreth. There are some hints that the Flood household was similarly shaped by paternal preoccupations with theories of individual rational action: his children were for example surprised when their father allowed them in this instance to cooperate rather than compete with one another.

2. Merrill M. Flood, "Some Experimental Games," RAND RM-789-1 (1952), 30.

3. On the early history of the RAND Corporation, see Bruce L. R. Smith, *The RAND Corporation: Case Study of a Nonprofit Advisory Corporation* (Cambridge, MA: Harvard University Press, 1966). See also chapter 3 in this book.

4. Flood, "Some Experimental Games," 1–4.

5. Olaf Helmer, "Strategic Gaming," RAND P-1902 (1960).

6. Robert F. Bales, Merrill M. Flood, and A. S. Householder, "Some Interaction Models," RAND RM-953 (1952), 26–42.

7. G. Kalisch et al., "Some Experimental N-Person Games," RAND RM-948 (1952), quoted in Robert Leonard, *Von Neumann, Morgenstern, and the Creation of Game Theory* (Cambridge: Cambridge University Press, 2010), 327. See also Leonard's account of the checkered career of game theory at RAND, 299–343.

8. Joseph Weizenbaum, *Computer Power and Human Reason: From Judgment to Calculation* (San Francisco: W. H. Freeman, 1976), 43.

9. Ibid., 223.

10. John von Neumann and Oskar Morgenstern, *Theory of Games and Economic Behavior*, 2nd ed. (Princeton, NJ: Princeton University Press, 1953), 33.

11. Jacob Marschak, "Rational Behavior, Uncertain Prospects, and Measurable Utilities," *Econometrica* 18 (1950): 112.

12. Ludwig Hoffmann, *Mathematisches Wörterbuch*, 7 vols. (Berlin: Wiegandt & Hempel, 1858–67).

13. Ivor Grattan-Guinness, *The Search for Mathematical Roots, 1870–1940: Logic, Set Theories, and the Foundations of Mathematics from Cantor through Russell to Gödel* (Princeton, NJ: Princeton University Press, 2000) traces the main developments. Jean Van Heijenoort, ed., *From Frege to Gödel: A Sourcebook in Mathematical Logic, 1879–1931* (Cambridge, MA: Harvard University Press, 1967), contains the key texts.

14. A. A. Markov, *Theory of Algorithms*, trans. Jacques J. Schorr-Kon and the PST staff (Moscow: Academy of Sciences of the USSR, published for the National Science Foundation and the Department of Commerce, USA by the Israel Program for Scientific Translation, 1954), 1.

15. Herbert Simon, *Models of Man: Social and Rational. Mathematical Essays on Rational Behavior in a Social Setting* (New York: Wiley, 1957), 202; see also Hunter Crowther-Heyck, *Herbert A. Simon: The Bounds of Reason in Modern America* (Baltimore, MD: Johns Hopkins University Press, 2005), 113–18.

16. Thomas C. Schelling, *The Strategy of Conflict* (1960; Cambridge, MA: Harvard University Press, 1980), 83–86.

17. "Cooperation-non-coop^with without side payments. The intuitive background: is cooperation more 'natural'? All evidence points to it: from any organiz[ation] of a firm (farm!) from primitive man on." Oskar Morgenstern, handwritten note, dated December 29, 1960, Oskar Morgenstern Papers, Box 42, "Game Theory Book: Notes and Papers, 1957–1969, n.d." folder, Duke University Special Collections, Durham, NC.

18. On the context of Leibniz's *characteristica universalis* as part of his plan for the *Demonstrationes Catholicae*, see Maria Rosa Antognazza, *Leibniz: An Intellectual Biography* (Cambridge: Cambridge University Press, 2009), 90–100; on Daniel Bernoulli's 1738 revisions of mathematical expectation as a solution to the St. Petersburg paradox, see Lorraine Daston, *Classical Probability in the Enlightenment* (Princeton, NJ: Princeton University Press, 1988), 70–77; on Babbage's analytical engine, see Simon Schaffer, "Babbage's Intelligence: Calculating Engines and the Factory System," *Critical Inquiry* 21 (1994): 203–27; and on Jevons's logical machines, see Harro Maas, *William Stanley Jevons and the Making of Modern Economics* (Cambridge: Cambridge University Press, 2005), 123–50.

19. L. F. Menabrea, "Sketch of the Analytical Engine Invented by Charles Babbage," trans. Ada Augusta, Countess of Lovelace, first published in the *Bibliothèque de Genève* in 1842 and reprinted in Philip Morrison and Emily Morrison, eds., *Charles Babbage and His Calculating Engines*, 225–95 (New York: Dover, 1961), 225. This account had Babbage's express approval: Charles Babbage, *Passages from the Life of a Philosopher* [1864], excerpts reprinted in Morrison and Morrison, *Charles Babbage and His Calculating Engines*, 64–68.

20. "The idea behind digital computers may be explained by saying that these machines are intended to carry out any operations which could be done by a human com-

puter. The human computer is supposed to be following fixed rules; he has no authority to deviate from them in any detail" (A. M. Turing, "Computing Machinery and Intelligence," *Mind* 59 (1950): on 436).

21. Wilson didn't know the source but consulted Ian Hacking, then visiting from Cambridge, England, who was able to supply it. Oskar Morgenstern to Margaret Wilson, December 14, 1971, with memo concerning Wilson's telephone call from January 6, 1972. Oskar Morgenstern Papers, Box 42, "Game Theory History II: Notes and Papers, 1971–73" folder, Duke University Special Collections, Durham, NC.

22. Gottfried Wilhelm Leibniz, "Preface to the General Science [1677]," in *Leibniz Selections*, ed. Philip Wiener, 12–17 (New York: Charles Scribner's Sons, 1951), 15.

23. On Morgenstern's early intellectual formation see Robert Leonard, "Between Worlds,' or an Imagined Reminiscence by Oskar Morgenstern about Equilibrium and Mathematics in the 1920s," *Journal of the History of Economic Thought* 26 (2004): 285–310, and Leonard, *Von Neumann, Morgenstern, and the Creation of Game Theory*, 71–184.

24. Morgenstern's arguments about the in-principle impossibility of economic forecasting were first set forth in his *Wirtschaftsprognose. Eine Untersuchung ihrer Voraussetzungen und Möglichkeiten* (Vienna: Julius Springer, 1928).

25. See, e.g., Kenneth J. Arrow, *Social Choice and Individual Values*, 2nd ed. (1953; New Haven, CT: Yale University Press, 1963), 93–96, concerning the Condorcet election paradoxes, and more generally, Duncan Black, *The Theory of Committees and Elections* (Cambridge: Cambridge University Press, 1958), 159–80.

26. M. J. A. N. Condorcet, *Essai sur l'application de l'analyse à la pluralité des décisions rendues à la pluralité des voix* (Paris: Imprimerie Royale, 1785), cxiii–iv.

27. Immanuel Kant, *Critique of Pure Reason* [1781, 1787], trans. and ed. Paul Guyer and Allen W. Wood (Cambridge: Cambridge University Press, 1997), A824–5/B852–3.

28. Jacques Vaucanson, *Le mécanisme du flûteur automate* (Paris: Guerin, 1738); see also Horst Bredekamp, *Antikensehnsucht und Maschinenglauben. Die Geschichte der Kunstkammer und die Zukunft der Kunstgeschichte* (Berlin: Klaus Wagenbach, 1993); Jessica Riskin, "The Defecating Duck, or the Ambiguous Origins of Artificial Life," *Critical Inquiry* 29 (2003): 599–633.

29. Julien Offray de la Mettrie, *Man a Machine* [1748], ed. Gertrude Carman Bussey, French–English edition (La Salle: Open Court, 1912), 93, 128.

30. Simon Schaffer, "Enlightened Automata," in *The Sciences in Enlightened Europe*, ed. William Clark, Jan Golinski, and Simon Schaffer, 126–65 (Chicago: University of Chicago Press, 1999), 156.

31. Turing, "Computing Machinery and Intelligence," 438.

32. Gottfried Wilhelm Leibniz, "Towards a Universal Characteristic [1677]," in Wiener, *Leibniz Selections*, 23.

33. Leibniz, "Preface," 16; Antognazza, *Leibniz*, 234–38.

34. These analogies between Condorcet's work and modern preoccupations have been pressed in Gilles-Gaston Granger, *La Mathématique sociale du marquis de Condorcet* (Paris: Presses universitaires de France, 1956); and Keith Michael Baker, *Condorcet: From Natural Philosophy to Social Mathematics* (Chicago: University of Chicago Press, 1975).

35. M. J. A. N. Condorcet, *Esquisse d'un tableau historique de progrès de l'esprit humain*, ed. O. H. Prior, édition présentée par Yvon Belaval (Paris, Librairie philosophique J. Vrin, 1970), 174.

36. Keith Michael Baker, "An Unpublished Essay by Condorcet on Technical Methods of Classification," *Annals of Science* 18 (1962): 104.

37. See the manuscript report, dated April 30, 1785, concerning a proposed prize to be offered by the Académie Royale des Sciences, Dossier Condorcet, Archives de l'Académie des Sciences, Paris.

38. See, e.g., Condorcet, *Essai*, lxxv.

39. Condorcet to the Count Pierre [Pietro] Verri, November 7, 1771, in F. Arago and A. Condorcet-O'Connor, eds., *Oeuvres de Condorcet*, vol. 1 (Paris: Firmin Didot Frères, 1847–49), 283–287. See Emma Rothschild, "Condorcet and the Conflict of Values," *Historical Journal* 3 (1996): 677–701.

40. M. J. A. N. Condorcet, *Vie de Turgot* [1786], in Arago and Condorcet-O'Connor, *Oeuvres de Condorcet*, vol. 5, 159–60.

41. M. J. A. N. Condorcet, *Élémens d'arithmétique et de géométrie* [1804], *Enfance* 4(1989): 44.

42. M. J. A. N. Condorcet, *Moyens d'apprendre à compter surement et avec facilité* (Paris: Moutardier, 1804), *Enfance* 4 (1989): 61–62.

43. Daston, *Classical Probability*, 49–111.

44. Joseph Hontheim, *Der logische Algorithmus in seinem Wesen, in seiner Anwendung und in seiner philosophischen Bedeutung* (Berlin: Felix L. Dames, 1895), 51.

45. Henry Thomas Colebrooke, "Address on Presenting the Gold Medal of the Astronomical Society to Charles Babbage," *Memoirs of the Astronomical Society* 1 (1852): 509. Despite the fortune expended on its manufacture, Babbage's plans for the difference engine were never fully realized in his lifetime: Doron Swade, *The Difference Engine: Charles Babbage and the Quest to Build the First Computer* (New York: Viking, 2001).

46. David Hilbert and Wilhelm Ackermann, *Grundzüge der theoretischen Logik* (Berlin: Springer, 1928), 77.

47. A. M. Turing, "On Computable Numbers, with an Application to the *Entscheidungsproblem*," *Proceedings of the London Mathematical Society*, ser. 2, 42 (1936–37): 253.

48. Philip Husbanks, Owen Holland, and Michael Wheeler, eds., *The Mechanical Mind in History* (Cambridge, MA: MIT Press, 2008); Pamela McCorduck, *Machines Who Think: A Personal Inquiry into the History and Prospects of Artificial Intelligence*, 2nd ed. (1979; Natick, MA: A.K. Peters, 2004), contains valuable interview material with AI pioneers.

49. See the relevant entries in Alain Rey, ed., *Le Robert Dictionnaire historique de la langue française*, 3 vols. (Paris: Dictionnaires Le Robert, 2000); *Oxford English Dictionary* (Oxford: Oxford University Press, 2010); Jacob and Wilhelm Grimm, *Deutsches Wörterbuch*, 33 vols. (Leipzig: S. Hirzel, 1893; facsimile reprint, 1991).

50. The centrality of the character and judgment of the abbot is emphasized from the very earliest commentaries: see, e.g., Smaragdus of Saint-Miel, *Commentary on the Rule of Saint Benedict*, trans. David Barry, with introductory essays by Terence Kardong, Jean Leclerq, and Daniel M. LaCorte (Kalamazoo, MI: Cistercian Publications, 2007), 18–23.

51. [Chevalier de Jaucourt], "REGLE, MODELE (*Synon.*)," in Denis Diderot and Jean d'Alembert, eds., *Encyclopédie, ou Dictionnaire raisonné des sciences, des arts et des métiers*, vol. 28 (Lausanne: Les sociétés typographiques, 1780), 116–17.

52. [Unsigned] "REGLE (*Dramatique, poésie*)," in Denis Diderot and Jean d'Alembert, eds., *Encyclopédie, ou Dictionnaire raisonné des sciences, des arts et des métiers*, vol. 28 (Lausanne: Les sociétés typographiques, 1780), 118; see also [Jean-François Marmontel], "REGLES (*Belles-lettres*)," in ibid., 127–30.

53. On Kant's use of the term "Regel," see Johannes Haag, "Regel," in *Kant-Lexikon*,

ed. Georg Mohr, Jürgen Stolzenberg, and Marcus Willaschek, 3 vols. (Berlin: De Gruyter, forthcoming). Kant deploys a rich vocabulary of terms, including the necessary "law," the contingent "rule," and the "subjective" maxim, to distinguish various kinds of regularities: see, e.g., Immanuel Kant, second introduction to *Critique of Judgment* [1790, 1793], trans. Werner S. Pluhar (Indianapolis: Hackett, 1987), Ak. 5.172–73, 11,12; Ak. 5.183–84, 23; Ak. 5.194, 34.

54. Kant, *Critique of Judgment*, I.46–47, Ak. 5.307–10, 174–78. On the impact of Kant's views on the impossibility of scientific genius, see Simon Schaffer, "Genius in Romantic Natural Philosophy," in *Romanticism in the Sciences*, ed. Andrew Cunningham and Nicholas Jardine, 82–98 (Cambridge: Cambridge University Press, 1990).

55. Kant distinguishes art as skill from science on the basis of whether theoretical knowledge alone (science) is or is not sufficient to bring about the desired effect: "[The Dutch anatomist Peter] Camper describes with great precision what the best shoe would have to be like, yet he was certainly unable to make one" (Kant, *Critique of Judgment*, I.43, Ak. 5.304, 171). An art, whether free or mercenary (i.e., a craft), cannot be mastered by explicit rules alone; otherwise it would be a science. On Kant's pragmatic, technical, and moral rules, see Thomas Sturm, *Kant und die Wissenschaften vom Menschen* (Paderborn: Mentis, 2009), 487–502.

56. There is no comprehensive history of how "mechanical" came to mean "automatic," but suggestive evidence for contrasting views of machines can be found in Otto Mayer, *Authority, Liberty, and Automatic Machinery in Early Modern Europe* (Baltimore, MD: Johns Hopkins University Press, 1986), and John Tresch, *The Romantic Machine: Utopian Science and Technology after Napoleon* (Chicago: University of Chicago Press, 2012).

57. On the word's Arabic–Spanish etymology, see "Algorisme," in Alain Rey, ed., *Le Robert Dictionnaire historique de la langue française*, 3 vols. (Paris: Dictionnaires Le Robert, 2000), 1:82.

58. Jean Marguin, *Histoire des instruments à calculer. Trois siècles de mécanique pensante 1642–1942* (Paris: Hermann, 1994). The first calculating machine to work reliably and be manufactured in large numbers was invented in 1821 by the French mathematician Charles Xavier Thomas de Colmar, who not coincidentally was the founder of several insurance companies that relied on massive actuarial calculations: R. Mehmke, "Numerisches Rechnen," in *Enzyklopädie der Mathematischen Wissenschaften*, 6 vols., ed. Wilhelm Franz Meyer (Leipzig: B.G. Teubner, 1898–1934), vol. 1, part 2, 959–78.

59. Gaspard Riche de Prony, *Notices sur les grandes tables logarithmiques et trigonométriques, adaptées au nouveau système decimal* (Paris: Firmin Didot, 1824), 5; Charles Babbage, "On the Division of Mental Labour [1832]," repr. Morrison and Morrison, *Charles Babbage*, 315–21.

60. Charles Babbage, *On the Economy of Machinery and Manufactures*, 4th ed. (London: Charles Knight, 1835), 201. Babbage's ideas on the division of labor seem to have been belied by the actual implementation of mixed skill-machinery regimes in manufacturing, at least in the American case: David A. Hounshell, *From the American System to Mass Production, 1800–1932* (Baltimore, MD: Johns Hopkins University Press, 1984), 33, 68–82.

61. Lorraine Daston, "Enlightenment Calculations," *Critical Inquiry* 21 (1994): 182–202; Schaffer, "Babbage's Intelligence." On women computers, see David Alan Grier, *When Computers Were Human* (Princeton, NJ: Princeton University Press, 2006); also M. Norton Wise, "The Gender of Automata in Victorian Science," in *Genesis Redux:*

Essays on the History and Philosophy of Artificial Life, ed. Jessica Riskin, 163–95 (Chicago: University of Chicago Press, 2007).

62. Edward Sang, "Remarks on the Great Logarithmic and Trigonometrical Tables Computed in the Bureau de Cadastre under the Direction of M. Prony," *Proceedings of the Royal Society of Edinburgh* (1874–75), 10; Theodore M. Porter, "Precision and Trust: Early Victorian Insurance and the Politics of Calculation," in *The Values of Precision*, ed. M. Norton Wise, 173–97 (Princeton, NJ: Princeton University Press, 1995).

63. This tendency was especially pronounced in operations research: see Judy L. Klein, introduction to "Protocols of War and the Mathematical Invasion of Policy Space, 1940–1960" (unpublished manuscript).

64. Charles Babbage, *The Ninth Bridgewater Treatise. A Fragment* (London: John Murray, 1837), 93–99.

65. Martin Davis, *The Universal Computer: The Road from Leibniz to Turing* (New York: W. W. Norton, 2000) tells the story in this fashion, as does (in a more anecdotal vein) David Berlinski, *The Advent of the Algorithm: The 300-Year Journey from an Idea to the Computer* (New York: Harcourt, 2000).

66. E. Roy Weintraub, ed., *Toward a History of Game Theory* (Durham, NC: Duke University Press, 1992); Philip Mirowski, "When Games Grow Deadly Serious: The Military Influence on the Evolution of Game Theory," in *Economics and National Security: A History of their Interaction*, ed. Craufurd D. Goodwin, 227–55 (Durham, NC: Duke University Press, 1991); Leonard, *Von Neumann, Morgenstern, and the Creation of Game Theory*.

67. Mina Rees, "The Mathematical Sciences and World War II," *American Mathematical Monthly* 87 (1980): 607–21; Fred M. Kaplan, *The Wizards of Armageddon* (1983; Stanford, CA: Stanford University Press, 1991); Alex Abella, *Soldiers of Reason: The Rand Corporation and the Rise of American Empire* (Orlando, FL: Harcourt, 2008).

68. R. Duncan Luce and Howard Raiffa, *Games and Decisions: Introduction and Critical Survey* (1957; Mineola, NY: Dover, 1985), 3, 10.

69. Edwards, *The Closed World*, 66–73; Sharon Ghamari-Tabrizi, *The Worlds of Herman Kahn* (Cambridge, MA: Harvard University Press, 2005), 171–74.

70. Robert J. Leonard, "Creating a Context for Game Theory," in Weintraub, *Toward a History of Game Theory*, 37, 48.

71. Angela M. O'Rand, "Mathematizing Social Science in the 1950s: The Early Development and Diffusion of Game Theory," in Weintraub, *Toward a History of Game Theory*, 189.

72. Luce and Raiffa, *Games and Decisions*, 1.

73. Ibid., 6. On the context and far-reaching implications of conceiving human behavior in terms of restricted choices, see Hunter Heyck, "Producing Reason," in *Cold War Social Science: Production, Liberal Democracy, and Human Nature*, ed. Mark Solovey and Hamilton Cravens, 99–116 (New York: Palgrave, 2012). We are grateful to Professor Heyck for allowing us to read his essay before publication.

74. Milestones would include Frege's *Begriffsschrift* (1879), Hilbert's *Entscheidungsproblem* (1927), and A. A. Markov's *Theory of Algorithms* (1954).

75. L. J. Savage, *Foundations of Statistics* (New York: John Wiley & Sons, 1954). This approach allowed Luce and Raiffa to abstract from the particulars of any given subjective utility curve, so long as preferences were transitive and consistent: Luce and Raiffa, *Games and Decisions*, 32.

76. Philip Mirowski, "What were von Neumann and Morgenstern Trying to Accomplish?" in Weintraub, *Toward a History of Game Theory*, 113–47. On the military links

to other formal theories of rationality, see also Robin E. Rider, "Operations Research and Game Theory," in Weintraub, *Toward a History of Game Theory*, 229–230; Edwards, *The Closed World*; and Amadae, *Rationalizing Capitalist Democracy*.

77. Luce and Raiffa, *Games and Decisions*, 96.

78. Cornelius M. Kerwin, *Rulemaking: How Government Agencies Write Law and Make Policy*, 3rd ed. (Washington DC: Congressional Quarterly Press, 2003), 9–15.

79. Alvin W. Gouldner, *Patterns of Industrial Bureaucracy* (Glencoe, IL: Free Press, 1954).

80. Robert Dubin, Review of Gouldner, *Patterns of Industrial Bureaucracy*, *American Sociological Review* 20 (1955): 122.

81. Herbert A. Simon, "Some Further Requirements of Bureaucratic Theory," in *Reader in Bureaucracy*, ed. Robert K. Merton, Alisa P. Gray, Barbara Hockey, and Hanan C. Selvin, 51–58 (Glencoe, IL: Free Press, 1952), 54.

82. For some idea of this literature, see the bibliographies in Robert Baldwin, *Rules and Government* (Oxford: Clarendon Press, 1995), and George A. Krause and Kenneth J. Meier, eds., *Politics, Policy, and Organizations: Frontiers in the Scientific Study of Bureaucracy* (Ann Arbor: University of Michigan Press, 2006). Barry Bozeman, *Bureaucracy and Red Tape* (Upper Saddle River, NJ: Prentice Hall, 2000) contains a glossary of the technical terms coined to describe various aspects of rule-governed bureaucracies (e.g., "rule drift," "rule strain," "rule sum") on pp. 185–86.

83. Paul E. Meehl, *Clinical versus Statistical Prediction. A Theoretical Analysis and a Review of the Literature* (Minneapolis: University of Minnesota Press, 1954); Meehl, "When Shall We Use Our Heads Instead of the Formula?" *Journal of Counseling Psychology* 4 (1957): 268–73; Meehl, "Causes and Effects of My Disturbing Little Book," *Journal of Personality Assessment* 50 (1986): 370–75.

84. Harold Garfinkel, "The Rational Properties of Scientific and Common Sense Activities," *Behavioral Sciences* 5 (1962): 81.

85. Robert F. Bales, "Social Interaction," RAND P-587 (1954), 7.

86. Sidney Verba, "Assumptions of Rationality and Non-Rationality in Models of the International System," *World Politics* 14 (1961): 98.

87. Herman Kahn, *On Thermonuclear War* (Princeton, NJ: Princeton University Press, 1960). On the responses and their context, see Margot A. Henriksen, *Dr. Strangelove's America: Society and Culture in the Atomic Age* (Berkeley: University of California Press, 1997); and Ghamari-Tabrizi, *The Worlds of Herman Kahn*.

88. Oskar Morgenstern to Philip J. Farley, March 1, 1971, Oskar Morgenstern Papers, Box 35, "Arms Control and Disarmament Agency, 1967–1971" folder, Duke University Special Collections, Durham, NC.

89. Alan Boring, "Computer System Reliability and Nuclear War," *Communications of the ACM* 30 (1987): 115.

90. RAND Corporation, *The RAND Corporation: The First Fifteen Years* (Santa Monica, CA: RAND Corporation, 1963), 30.

91. Anatol Rapoport and Carol Orwant, "Experimental Games: A Review," *Behavioral Science* 7 (1962): 1–37.

92. See, e.g., M. G. Weiner, "An Introduction to War Games," RAND P-1773 (1959); Olaf Helmer, "Strategic Games," RAND P-1902 (1960); Robert F. Bales, "Social Interaction," RAND P-587 (1954); see also chapter 4 in this volume.

93. Thomas C. Schelling, *The Strategy of Conflict* (1960; Cambridge, MA: Harvard University Press, 1980), 163.

94. Among the most influential were Hubert Dreyfus, *What Computers Can't Do: A Critique of Artificial Reason* (New York: Harper and Row, 1972); and Weizenbaum, *Com-*

puter Power. See also H. M. Collins, *Artificial Experts: Social Knowledge and Intelligent Machines* (Cambridge, MA: MIT Press, 1990), for a clear and comprehensive review of the arguments. The arguments of the late Wittgenstein about rule following cast a long shadow over these critiques: Ludwig Wittgenstein, *Philosophical Investigations* [1953], trans. G. E. M. Anscombe, 3rd ed. (Englewood Cliffs, NJ: Prentice Hall, 1958), §199, p. 81; §193, p. 78; cp. §219, p. 85, on the analogous symbolism of a rule as an infinite set of tracks determining the unlimited application of a rule.

95. See Luce and Raiffa, *Games and Decisions*, 12–38; Kenneth Arrow, *Social Choice and Individual Values*, 2nd ed. (1953; New Haven, CT: Yale University Press, 1963), 9–21.

96. Arrow, *Social Choice*, 60, 98; S. M. Amadae, *Rationalizing Capitalist Democracy: The Cold War Origins of Rational Choice Liberalism* (Chicago: University of Chicago Press, 2003), 102–16.

97. Schelling, *The Strategy of Conflict*, 4, 17.

98. Ibid., 12–13. Mohammed Mossadeq was the Iranian prime minister who tried to nationalize American and British oil holdings and who was violently deposed in 1953 with the connivance of the CIA and MI6.

99. Jon Elster, *Ulysses and the Sirens: Studies in Rationality and Irrationality* (1979; Cambridge: Cambridge University Press, 1990) remains the standard work.

100. Thomas C. Schelling, *Strategies of Commitment and Other Essays* (Cambridge, MA: Harvard University Press, 2006), 106–7. In the 1980s Schelling admitted that "it is only in economics that the individual is modeled as a coherent set of preferences and certain cognitive facilities" and called for modifications of rational choice theory to deal with the divided self (ibid., 74).

CHAPTER TWO

1. Lt. Fred McAfee, quoted in Paul Fisher, "The Berlin Airlift," *The Beehive, United Aircraft Corporation* 23 (1948): 14–15.

2. The Berlin Crisis, 1948, US Department of State, Foreign Policy Studies Branch, Division of Historical Policy Research, Research Project No. 171. Washington DC, 5.

3. The French also had a few planes helping out, but US and British planes carried most of the material for the French zone. At one point there was a dispute between the American air crews and the French when the former balked at hauling wine in. The outraged French "sent a delegation armed with their dietary history through all times. Their chief contention was that wine was to them equally as important as potatoes to a German, black bread to a Russian, or ketchup to a Texan." (Fisher, "The Berlin Airlift," 9.)

4. William H. Tunner, *Over the Hump* (1964; repr., Washington: Office of Air Force History, United States Air Force, 1985), 167. Tunner, a 1928 graduate of the US Military Academy at West Point, had during World War II successfully commanded the Hump operation that airlifted supplies from India over the Himalayas for the fighting in China. Tunner described his Asian, Berlin, and subsequent Korean airlift experiences in his memoirs.

5. Ibid., 174.

6. Ibid., 187.

7. The term "programming" in the "linear programming" and "dynamic programming" frameworks developed in the late 1940s draws from the traditional military use of the term as a time-phased scheduling of operations. Although the USAF Project SCOOP team designed algorithms to solve mathematical programming models

with digital computers, they used the term "coding," not programming, to describe the writing of machine instructions for the computer.

8. This history uses the terms "management science" and "operations research" interchangeably, but also acknowledges that in the early 1950s, some of the protagonists in our story elected to distinguish the two through separate professional organizations. Operations research emerged from the quantitative decision making in World War II brought to bear on the planning of specific military operations or the evaluation of alternative weapons systems. Phillip Morse and other World War II military operation researchers founded the Operations Research Society of America (ORSA) in 1952. Merrill Flood, as well as William Cooper, and Abraham Charnes from the Carnegie Institute, were key organizers of The Institute of Management Sciences (TIMS), founded in 1953; Project SCOOP's chief mathematician, George Dantzig, was a founding member of TIMS and his SCOOP colleague Murray Geisler served as eighth president. In the early 1950s ORSA had a stronger association with the military and concrete problem-orientated applications, and TIMS with the identification of basic research relevant to the practice of management. A common analogy of the distinction between operations research and management science was that of chemical engineering and chemistry respectively. Even in the early years, however, there was some overlap in membership, executive officers, and professional journal topics. Also the US military underwrote a considerable amount of the research that was published in both professional journals. ORSA and TIMS began sponsoring joint meetings in 1974 and formally merged in the Institute of Operations Research and Management Science in 1995. The early history of the two organizations is documented in Saul Gass and Arjang Assad, *Annotated Timeline of Operations Research* (New York: Kluwer Academic Publishers, 2005) and Gerald William Thomas, "A Veteran Science: Operations Research and Anglo-American Scientific Cultures, 1940–1960" (PhD diss., Harvard University, 2007). Merrill M. Flood, "The Objectives of TIMS," *Management Science* 2 (1956): 179; and Melvin Salvesen, "The Institute of Management Sciences: A Prehistory and Commentary on the Occasion of TIMS' 40th Anniversary," *Interfaces* 27, no. 3 (May–June 1997): 74–85, also shed light on the ORSA/TIMS distinctions.

9. George Dantzig (1914–2005) received his PhD in mathematics in 1946 from the University of Berkeley after working as a statistician first at the Bureau of Labor Statistics (1938–1939) and then at the Pentagon (1941–1946). Dantzig's solution to two unproven theorems in statistics, which he mistakenly assumed were homework assignments from his professor Jerzy Neyman, formed part of the story line in the film *Good Will Hunting*. Dantzig was a founding member of TIMS, its president in 1966, the first recipient in 1974 of the Von Neumann Theory Prize awarded jointly by the Operations Research Society of America and TIMS, and a recipient of the US National Medal of Science.

10. George B. Dantzig, "Concepts and Origins of Linear Programming," RAND P-980 (1957).

11. General Hoyt S. Vandenberg, "Air Force Letter No. 170–3, Comptroller Project SCOOP," Washington DC, October 13, 1948, Air Force Historical Research Agency IRIS Number 01108313: 1.

12. In his reflections on his experience with Project SCOOP, Lyle R. Johnson, "Coming to Grips with Univac," *IEEE Annals of the History of Computing Archive* 28 (2006): 42, described the high level of concern in 1948 over a possible World War III, which prompted the Air Force to systematically study resources for rapid mobilization.

13. The US monopoly lasted until the USSR successfully tested their first atomic bomb on August 29, 1949. Michael Gordin, *Red Cloud at Dawn: Truman, Stalin, and the End of the Atomic Monopoly* (New York: Farrar, Straus and Giroux, 2009), explores US military and diplomatic strategy during the four-year monopoly.

14. In 1947, Dantzig would describe what he was doing as "programming in a linear structure." In an informal encounter at the July 1948 RAND Corporation colloquium "Theory of Planning," Tjalling Koopmans suggested that Dantzig shorten the name to "linear programming." George Dantzig, "Linear Programming," in *History of Mathematical Programming: A Collection of Personal Reminiscences*, ed. Jan Lenstra, Alexander Kan, and Alexander Schrijver (Amsterdam: CWI, 1991), 29.

15. In *Linear Programming and Extensions* (Princeton, NJ: Princeton University Press, 1963), 16–18, George Dantzig discusses the inspiration to his own work of Leontief's quantitative model and the Bureau of Labor Statistics' use of input/output matrices of interindustry data during World War II. See also Martin C. Kohli, "Leontief and the U.S. Bureau of Labor Statistics, 1941–1954: Developing a Framework for Measurement," in *The Age of Economic Measurement*, ed. Judy L. Klein and Mary S. Morgan, 190–212 (Durham, NC: Duke University Press 2001); and Judy Klein, "Reflections from the Age of Measurement," in Klein and Morgan, *The Age of Economic Measurement*, 128–33.

16. Marshall K. Wood and Murray A. Geisler, *Machine Computation of Peacetime Program Objectives and Mobilization Programs*, Project SCOOP No. 8, report prepared for Planning Research Division Director of Program Standards and Cost Control, Comptroller, Headquarters US Air Force (Washington DC, July 18, 1949), 36.

17. Alfred Cowles, president of an investment firm in Colorado Springs, had a keen Depression-honed interest in the accuracy of stock market forecasts. In 1932, he established the Cowles Commission for Research in Economics. From the outset the commission had close ties with the Econometric Society and supported the statistical and mathematical research of prominent economists. In 1939, the Cowles Commission moved to the University of Chicago. In 1948, Koopmans took over as director of research and increased the commission's emphasis on mathematical methods for the study of rational behavior. Dantzig discusses his June 1947 meeting with Koopmans at the Cowles Commission in Chicago in "Linear Programming."

18. The simplex algorithm relies on the geometric property that the objective function will have a maximum value at a corner (vertex) of the convex feasible region bounded by the linear inequality constraints of the problem (e.g., arising from resource limitations or technological constraints). Dantzig's algorithm was an iterative method for moving about the geometric form (created by the constraints) to find the point where the objective function was at its maximum. Acknowledging the efficiency and widespread successful application of the simplex algorithm, the journal *Computing in Science & Engineering* named it one of the ten algorithms with the greatest influence in the twentieth century (John C. Nash, "The (Dantzig) simplex method for linear programming," *Computing in Science and Engineering* 2, no. 1 [2000]: 29–31).

19. US Air Force Planning Research Division Director of Program Standards and Cost Control Comptroller, *Scientific Planning Techniques: A Special Briefing for the Air Staff 5 August 1948*, Project SCOOP Discussion Papers, 1-DU, Washington DC, 10. The conference that Dantzig referred to is most likely the month-long colloquium on "Theory of Planning" that the RAND Corporation held simultaneously with a colloquium on game theory during July 1948. Paul Erickson, "Optimism and Optimi-

zation," chapter 3 in Erickson, "The World the Game Theorists Made" (unpublished manuscript, August 2012, MS Word files), has documented the stellar list of thirty-eight mathematicians, economists, and operations researchers, including Dantzig, Wood, and Koopmans, who participated in the planning colloquium and the enduring effect of the simultaneous colloquia on "cementing a relationship between game theory, programming, and the needs of the Air Force."

20. Tjalling Koopmans autobiography, http://nobelprize.org/nobel_prizes/economics/laureates/1975/koopmans.html (accessed August 6, 2011); Tjalling Koopmans, "Concepts of Optimality and Their Uses," http://nobelprize.org/nobel_prizes/economics/laureates/1975/koopmans-lecture.pdf (accessed August 6, 2011).

21. Cowles Commission, "Report for Period January 1, 1948–June 30, 1949," http://cowles.econ.yale.edu/P/reports/1948-49.htm (accessed August 6, 2011). The new focus on welfare economics and optimal behavior also corresponded with a change in 1948 in the research directorship at Cowles from Jacob Marschak to Tjalling Koopmans. That transition and the influence of military patronage at the Cowles Commission (with, e.g., in 1951 the RAND Corporation underwriting 32 percent of the budget and the Office of Naval Research 24 percent) is documented in Philip Mirowski, *Machine Dreams: Economics becomes a Cyborg Science* (Cambridge: Cambridge University Press, 2002), 215–22.

22. Donald Albers and Constance Reid, "An Interview with George B. Dantzig: The Father of Linear Programming," *College Mathematics Journal* 17, no.4 (1986): 309.

23. David Gale, Harold W. Kuhn, and Albert W. Tucker, "Linear Programming and the Theory of Games," in *Activity Analysis of Production and Allocation: Proceedings of a Conference*, Cowles Commission for Research in Economics Monograph No. 13, ed. Tjalling C. Koopmans (New York: John Wiley & Sons, 1951), 317–29; and George B. Dantzig, "A Proof of the Equivalence of the Programming Problem and the Game Problem," in Koopmans, *Activity Analysis of Production and Allocation*, 330–38. In his chapter on "Optimism and Optimization," Erickson demonstrates how the recognition in 1947 of the essential connection between linear programming and the two-person zero-sum game spurred the development of game theory at the RAND Corporation, where it became strongly linked to the study of warfare through optimization, and Princeton University (professional home of Kuhn and Tucker), where the two mathematical approaches were used for Office of Naval Research projects on logistics. As game theory strayed outside the sheltering matrix of the zero-sum game (see chapter 5), its clear connection with optimization was severed and mathematical programming began to trump game theory as the research foci of RAND mathematicians.

24. In 1948, Dantzig (US Air Force, *Scientific Planning Techniques*, 14) explained to the air staff that their existing punch-card calculators could only do 1 multiplication every two seconds, compared with the 1,000 multiplications per second in the electronic digital computers Project SCOOP was designing. The UNIVAC that the air force eventually acquired in 1952 could do about 465 multiplications per second and the SEAC, which the air force had begun using in 1951, had a capacity for 330 multiplications per second .

25. Under a World War II contract with the US Army, Eckert and Mauchly designed the ENIAC for ballistics research and delivered it to the Aberdeen Proving Ground in 1946. By 1948, the ENIAC was still the only electronic, large-scale digital computer that was in operation in the United States (there were five digital electromechanical relay computers in operation and another eight of various types under

development). The US Census Bureau had agreed to use the first UNIVAC built in the factory setting, but the air force was the first institution to take delivery of a UNIVAC, having to reassemble five thousand vacuum tubes. The details and significance of the pioneering efforts by the USAF to support development of electronic digital computers are documented in US Air Force, *Scientific Planning Techniques*, 13–15; and by Johnson, "Coming to Grips with Univac." Other discussions of the early computational history of linear programming include Edward Dunaway, US Air Force Oral History Interview by Daniel R. Mortensen, April 17, 1980, Transcript, Office of Air Force History IRIS No. 01129703; Edward Dunaway, US Air Force Oral History Interview by James R. Luntzel, June 22, 1973, Transcript, Office of Air Force History IRIS No. 01129703; Saul I. Gass, "Model World: In the Beginning There Was Linear Programming," *Interfaces* 20 (1990): 128–32; William Orchard-Hays, "Evolution of Linear Programming Computing Techniques," *Management Science* 4 (1958): 183–90; William Orchard-Hays, "History of the Development of LP Solvers," *Interfaces* 20 (1990): 61–73; Alex Orden, "LP from the '40s to the '90s," *Interfaces* 23 (1993): 2–12; and Emil D. Schell, "Application of the Univac to Air Force Programming," *Proceedings of the Fourth Annual Logistics Conference* (Navy Logistics Research Project, Washington DC, 1953): 1–7.

26. George B. Dantzig, "The Diet Problem," *Interfaces* 20 (1990): 43–47; Dantzig, *Linear Programming and Extensions*, 551; George Stigler, "The Costs of Subsistence," *Journal of Farm Economics* 27, no.2 (May 1945): 303–14.

27. Dantzig, "The Diet Problem"; Mina Rees, "The Mathematical Sciences and World War II," *American Mathematical Monthly* 87, no.8 (October 1980): 618.

28. Dantzig described the long process of exploring the computational efficiency of the simplex algorithm in his article "Impact of Linear Programming on Computer Development" (*OR/MS Today* 15 [August 1988]: 12–17), and in his interview with Donald Albers and Constance Reid in 1986.

29. Wood and Geisler, *Machine Computation*, 49.

30. Wood, "Research Program at Project SCOOP," *Symposium on Linear Inequalities and Programming Washington DC, June 14–16, 1951*, Project SCOOP Manual No. 10 (April 1, 1952), 7.

31. Wood and Geisler, *Machine Computation*, 7.

32. Ibid., 2.

33. Marshall K. Wood and George B. Dantzig, "Programming of Interdependent Activities: I General Discussion," *Econometrica* 17, no. 3/4 (Jul.–Oct. 1949): 198; and George Dantzig, "Programming of Interdependent Activities: II Mathematical Model," *Econometrica* 17, No. 3/4 (Jul.–Oct. 1949): 200–11.

34. Tjalling C. Koopmans, ed., *Activity Analysis of Production and Allocation: Proceedings of a Conference*, Cowles Commission for Research in Economics Monograph no. 13 (New York: John Wiley & Sons, 1951).

35. Marshall K. Wood and Murray A. Geisler, "Development of Dynamic Models for Program Planning," in Koopmans, *Activity Analysis of Production and Allocation*, 194 (original emphasis).

36. In his 1947 discussion on the benefits of the linear programming approach over his own World War II "program for programing," Dr. Edmund Learned explained to the air staff:

> The most vital contribution which this new processing technique—new analytical technique—makes to the Air Staff is in the area of alternatives. As all of you know, we had plenty of problems during the war guessing at the general pattern

of a program then running out the programing details, finding our lacks of balance and bottlenecks after we had made the run . . . with this speedier method of computation it is possible for a Staff officer or the Chief of Staff to define a number of alternatives that he would like to see worked out in detail. They can be run rapidly on the machine. (US Air Force, *Scientific Planning Techniques*, 26)

37. Minor-scale optimization was still possible and could be expanded with the expansion of computing capacity. As Wood and Geisler explained to the air staff:

the triangular procedure has been organized so that small local maximization problems can readily be introduced into local areas of the model where alternatives are to be considered. As equipment of greater computing capacity becomes available, these areas can be gradually enlarged, permitting more and more consideration of alternatives. Thus the transition from a determinate model permitting only a single solution to an indeterminate model in which we select the best from among alternative solutions will be gradual, rather than an abrupt one. (Wood and Geisler, *Machine Computation*, 48)

38. Wood and Geisler, "Development of Dynamic Models," 206.
39. Murray A. Geisler, *A Personal History of Logistics* (Bethesda, MD: Logistics Management Institute, 1986), 6.
40. Ludwig von Mises, "Economic Calculation in the Socialist Commonwealth" (1920), repr. in *Collectivist Economic Planning; Critical Studies on the Possibilities of Socialism*, ed. Friedrich. A. Hayek (London: Routledge & Kegan Paul, 1935), 105.
41. Tjalling C. Koopmans, Introduction to Koopmans, *Activity Analysis of Production and Allocation*, 7.
42. E.g., Stephen Enke, an economist at the RAND Corporation, demonstrated that with linear programming economists could contribute "the principles of value determination and the logic of economizing" to determine the production, consumption, and allocation of the fissionable materials U235 and Pu239 for which there were no markets or real prices (Enke, "Some Economic Aspects of Fissionable Material," *Quarterly Journal of Economics* 68, no. 2 [1954]: 217).
43. Cowles Commission for Research in Economics, *Rational Decision-Making and Economic Behavior, 19th Annual Report, July 1, 1950–June 30, 1951*, http://cowles.econ .yale.edu/P/reports/1950–51.htm (accessed August 6, 2011).
44. Economists use the term "normative" to mean prescriptive—declaring what ought to be. This is in contrast to "positive" economics, which describes what is.
45. Herbert Simon (1916–2001) described himself a mathematical social scientist. As an undergraduate at the University of Chicago and a graduate student at the University of California at Berkeley, Simon sought out courses in physics, mathematical economics, and symbolic logic. Simon's doctoral thesis on administrative decision making built on his operations research work for a city government. From 1949 until his death, Simon was a professor at the Carnegie Institute of Technology/Carnegie Mellon University. He received the Nobel Prize in Economics in 1978. The broad disciplinary span of Simon's work, including administrative decision making and computer-simulated problem solving, and his departmental travels at Carnegie Mellon through industrial administration, psychology, and political science, are addressed in Hunter Crowther-Heyck, *Herbert A. Simon: The Bounds of Reason in Modern America*, (Baltimore, MD: Johns Hopkins University Press, 2005); Esther Mirjam Sent, "Herbert A. Simon as a Cyborg Scientist," *Perspectives on Science* 8, no.4 (2000): 380–406; and Sent, "Simplifying Herbert Simon," *History of Political Economy* 37, no. 2 (2005): 227–32. In *Models of a Man*, edited by Mie Augier and

James March (Cambridge, MA: MIT Press, 2004), forty of Simon's former colleagues and research partners, reflect on Simon's contributions to social science. In *Machine Dreams*, Philip Mirowski examines Simon's research for the Cowles Commission, 452–72.

46. Simon's 1952 notes on the GSIA "Research Budget 1950–1952," (box 18, folder 1214, Herbert A. Simon Collection, Carnegie Mellon University Archives, Pittsburgh, PA) indicate that the contract with the air force and the Bureau of the Budget financed on an annual basis the equivalent of three man-years of faculty research and six man-years of graduate assistant research per year in addition to overhead costs.

47. Charles C. Holt, "Servo Control and Programming—Polar Philosophies of Optimizing Behavior," report November 20, 1951, folder GSIA—Air Force Research Project #6, Herbert A. Simon Collection, Carnegie Mellon University Archives, Pittsburgh, PA.

48. Abraham Charnes, William. W. Cooper, and B. Mellon, "Blending Aviation Gasolines—A Study in Programming Interdependent Activities." Paper read at Project SCOOP Symposium on Linear Inequalities and Programming, June 14–16, 1951.

49. Abraham Charnes, "Optimality and Degeneracy in Linear Programming," *Econometrica* 20, no.2 (1952): 169–70.

50. William W. Cooper, Abraham Charnes, and A. Henderson, *An Introduction to Linear Programming* (New York: Wiley, 1953).

51. Hunter Crowther-Heyck (*Herbert A. Simon*, 184–214) examines the abiding theme of adaptation in Simon's models and its foundation on servo theory. Servomechanisms are error-actuated devices whose output controlled a system in response to the information, provided through a feedback loop, on the difference between the actual state and the desired (or predicted) state.

52. Herbert A. Simon, "Notes on Two Approaches to the Production Rate Problem," (Cowles Commission Discussion Paper: Economics No. 2057, November 19, 1952), 2.

53. The status of the air force research projects and the transfer of these to the ONR is discussed in the GSIA's final report to Project SCOOP (William Cooper, *Final Report on Intra-Firm Planning and Behavior: A Research Project Sponsored by the U. S. Department of the Air Force at the Carnegie Institute of Technology Graduate School of Industrial Administration*, July 1, 1953, box 15, folder 1072, Herbert A. Simon Collection, Carnegie Mellon University Archives, Pittsburgh, PA. It was a relatively seamless transfer partly because as early as May 1948 Dantzig had briefed the Navy Staff on the aims and methods of Project SCOOP and had maintained contact in the course of developing the wartime and peacetime programs. Judy L. Klein, *Protocols of War and the Mathematical Invasion of Policy Space, 1940–1970* (unpublished manuscript, August 2012, MS Word), examines the application-driven theory that came out of both GSIA military planning contracts, including new forecasting tools, the articulated significance of quadratic cost functions for deriving linear decision rules, rational expectations, and the technical revolution in economic theory initiated by a strong emphasis on modeling strategies.

54. During this time, Simon's research was also funded by a RAND Corporation contract with the Cowles Commission on the theory of resource allocation and an ONR contract with Cowles for research on decision making under uncertainty.

55. Impressed with Dantzig's demonstration of how to achieve effective numerical solutions at the 1948 colloquium "Theory of Planning," Richard Bellman, a mathematician at the RAND Corporation, developed a new form of optimization for multi-

stage decision processes. Bellman's dynamic programming included an economic criterion function maximizing damage to the enemy or minimizing losses and was essentially an algorithm for determining the optimal allocation of resources over time in contrast with linear programming's optimal allocation of resources across activities. Bellman first developed the protocol to give a rule at each stage of a nuclear war for deciding which enemy target to bomb with the relatively scarce atomic weapons. (Judy L. Klein, *Protocols of War*.) As with linear programming, the optimization tool was quickly appropriated to other contexts via the academic-military-industrial complex exemplified by military-funded research at the Carnegie Institute.

56. Herbert A. Simon, "Dynamic Programming under Uncertainty with a Quadratic Criterion Function," *Econometrica* 24 (1956): 74–81.

57. Charles C. Holt, "Superposition Decision Rules for Production and Inventory Control," ONR Research Memorandum No. 3, October 27, 1953, box 18, folder 1221, Herbert A. Simon Collection, Carnegie Mellon University Archives, Pittsburgh, PA.

58. This was not the first time Simon challenged economists and their discipline. In *Herbert A. Simon*, Hunter Crowther-Heyck documents the tensions spanning decades between Simon and his Carnegie colleagues in economics. Nor was this the first time that social scientists, or Simon himself, seriously contemplated notions of limited rationality. In their essay "The Conceptual History of the Emergence of Bounded Rationality" (*History of Political Economy* 37, no. 1 [2005]: 27–59), Matthias Klaes and Esther-Mirjam Sent construct a conceptual trajectory of first time occurrences of a family of expressions of limited, approximate, incomplete, and bounded rationality.

59. Herbert A. Simon, "A Behavioral Model of Rational Choice," *Quarterly Journal of Economics* 69, no. 1 (1955): 99.

60. Herbert A. Simon, *Models of Man: Social and Rational: Mathematical Essays on Rational Human Behavior in a Social Setting* (New York: Wiley, 1957), 204. Simon used the Scottish/Northumbrian term of "satisfice" meaning "satisfy." In his Nobel Prize lecture in Stockholm on December 8, 1978, Simon reflected on the conceptual development of bounded rationality by describing the Carnegie team's quadratic cost approximation to illustrate "how model building in normative economics is shaped by computational considerations" (Herbert A. Simon, "Rational Decision Making in Business Organizations." *American Economic Review* 69 [1979]: 498).

61. Herbert A. Simon, "Theories of Bounded Rationality," in *Decision and Organization*, ed. C. B. McGuire and Roy Radner, 161–76 (Amsterdam: North-Holland, 1972).

62. Herbert A. Simon, "From Substantive to Procedural Rationality," lecture at Groningen University, September 24, 1973, box 81, folder 6519, Herbert A. Simon Collection, Carnegie Mellon University, Pittsburgh, PA; "From Substantive to Procedural Rationality," in *Method and Appraisal in Economics*, ed. S. J. Latsis, 120–48 (New York, 1976). In many of Simon's notes and publications, starting with his dissertation in 1943, the contrasting adjectives of "substantial" and "procedural" appear in the same paragraph. Over the decades Simon paired the following nouns with the two adjectives: "matters," "conformity," "decision premises," "flexibility," "problems," and "alternatives." In his 1964 essay "Rationality" (*Dictionary of the Social Sciences*, edited by J. Gould and W.L. Kolb (Glencoe, IL: Free Press, 1964), 574), Simon contrasted two types of rationality, the economist's "attribute of an action selected by a choice process" and the psychologist's "processes of choice that employ the intellectual faculty." It was apparently not until 1973, however, that Simon coined the phrases "substantial rationality" and "procedural rationality" and subsequently

highlighted the distinction in several key publications: See, e.g., Herbert A. Simon, "On How to Decide What to Do," *Bell Journal of Economics* 9 (1978): 494–507; Simon, "Rationality as Process and as Product of Thought," *American Economic Review* 68 (1978): 1–16; and Simon, "Rational Decision Making in Business Organizations," *American Economic Review* 69 (1979): 493–513.

63. That search is ongoing, see William J. Cook, *The Pursuit of the Traveling Salesman: Mathematics at the Limits of Computation* (Princeton, NJ: Princeton University Press, 2012).

64. Simon, "From Substantive to Procedural Rationality," 140.

65. Simon, "Theories of Bounded Rationality," 167.

66. "Economics and Operations Research: A Symposium," *Review of Economics and Statistics* 40, no. 3 (1958): 195–229.

67. Wilson's antipathy to planning and its effect on the disbandment of Project SCOOP is discussed in Lyle R. Johnson, "Coming to Grips with Univac," 42; and by Edward Dunaway in his 1980 US Air Force Oral History Interview, 5. In his *Personal History of Logistics*, Murray Geisler remarked, "We learned how fragile a research group can be. . . . A large bureaucracy does not value such groups adequately; they are doomed to have a limited life"(11).

68. Walter Jacobs, "Air Force Progress in Logistics Planning," *Management Science* 3 (1957): 213–24.

69. Gass described a typical aircraft deployment linear programming problem that he worked on in the mid-1950s, "Given the initial availability of a combat-type aircraft and the additional monthly availabilities in the succeeding months, divide these availabilities between combat and training so as to maximize, in some sense, the combat activity" (Saul I. Gass, "Model World," 131).

70. *Management Science* published an updated, expanded English translation of Kantorovich's 1939 booklet in July 1960 ("Mathematical Methods of Organizing and Planning Production," *Management Science* 6, no. 4 [1960]: 366–422). In his foreword, Tjalling Koopmans described the study as "an early classic in the science of management under any economic system" ("A Note about Kantorovich's Paper, 'Mathematical Methods of Organizing and Planning Production,'" *Management Science* 6 no. 4 [1960]: 364). Johanna Bockman and Michael A. Bernstein discuss the correspondence and Cold War mediated relationship between Kantorovich and Koopmans in "Scientific Community in a Divided World: Economists, Planning, and Research Priority during the Cold War," *Comparative Studies in Society and History* 50, no. 3 (2008): 581–613.

71. William Orchard-Hays, "History of the Development of LP Solvers," *Interfaces* 20 (1990): 62.

72. Herbert A. Simon, "On How to Decide," 496.

73. Ellsberg is most well known for his 1971 leak of *The Pentagon Papers*, top-secret decision-making documents on US involvement in Vietnam. Before that he had served as a nuclear strategist at the RAND Corporation and under Secretary of Defense Robert McNamara. While at RAND, Ellsberg demonstrated through experiments a paradox in decision making that violated the assumptions of the expected utility hypothesis so critical to the game theoretic perspective we will encounter in chapter 5 ("Risk, Ambiguity, and the Savage Axioms," *Quarterly Journal of Economics* 75, no. 4 [1961]: 643–69). Additional experiments that called into question the strict assumptions of economic rationality will be explored in chapter 6.

Notes to Pages 79–90 / 209

Notes to Pages 79–90 / 209

74. Daniel Ellsberg, "A Final Comment," *Review of Economics and Statistics* 40, no. 3 (1958): 229.

75. Dantzig, "Linear Programming."

76. George B. Dantzig, "Management Science in the World of Today and Tomorrow," *Management Science* 13 (1967): C-109 (original emphasis).

CHAPTER THREE

1. The classic eyewitness account, which is the base of all later interpretations of the crisis, is Robert F. Kennedy, *Thirteen Days: A Memoir of the Cuban Missile Crisis* (1968; New York: W. W. Norton, 1971).

2. Sidney Verba, "Assumptions of Rationality and Non-Rationality in Models of the International System," *World Politics* 14 (1961): 95 (ellipses added).

3. Richard Ned Lebow, "The Cuban Missile Crisis: Reading the Lessons Correctly," *Political Science Quarterly* 98 (1983): 458. For an emphasis on "rationality," see Graham T. Allison, *Essence of Decision: Explaining the Cuban Missile Crisis* (Boston: Little, Brown, 1971).

4. George F. Kennan, *The Nuclear Delusion: Soviet-American Relations in the Atomic Age* (New York: Pantheon, 1983), 176.

5. On the history of nuclear strategy, see Lawrence Freedman, *The Evolution of Nuclear Strategy*, 3rd ed. (New York: Palgrave Macmillan, 2003); and Fred Kaplan, *The Wizards of Armageddon* (New York: Simon and Schuster, 1983).

6. David Alan Rosenberg, "The Origins of Overkill: Nuclear Weapons and American Strategy, 1945–1960," *International Security* 7 (1983): 3–71; Michael D. Gordin, *Red Cloud at Dawn: Truman, Stalin, and the End of the Atomic Monopoly* (New York: Farrar, Straus & Giroux, 2009).

7. Bertrand Russell, *Common Sense and Nuclear Warfare* (London: George Allen & Unwin, 1959), 30.

8. Herman Kahn, *On Thermonuclear War*, 2nd ed. (Princeton, NJ: Princeton University Press, 1961), 165. On the idiosyncracy of Kahn's stance, see Freedman, *The Evolution of Nuclear Strategy*, 204–5.

9. Anatol Rapoport, "Chicken à la Kahn," *Virginia Quarterly Review* 41 (1965): 370–89.

10. For a full treatment of Kahn, see Sharon Ghamari-Tabrizi, *The Worlds of Herman Kahn: The Intuitive Science of Thermonuclear War* (Cambridge, MA: Harvard University Press, 2005).

11. Herman Kahn, *On Escalation: Metaphors and Scenarios* (New York: Frederick A. Praeger, 1965), 10–12.

12. Quoted in ibid., 226.

13. Quoted in Kennedy, *Thirteen Days*, 67–68.

14. Elizabeth Converse, "The War of All against All: A Review of *The Journal of Conflict Resolution*, 1957–1968," *Journal of Conflict Resolution* 12 (1968): 471–532; Martha Harty and John Modell, "The First Conflict Resolution Movement, 1956–1971: An Attempt to Institutionalize Applied Interdisciplinary Social Science," *Journal of Conflict Resolution* 35 (1991): 720–58. The journal later moved to Yale and became primarily a venue for game theory.

15. Svenn Lindskold, "Trust Development, the GRIT Proposal, and the Effects of Conciliatory Acts on Conflict and Cooperation," *Psychological Bulletin* 85 (1978): 778.

16. Charles E. Osgood, "Suggestions for Winning the Real War with Communism." *Journal of Conflict Resolution* 3 (1959): 315 (original emphasis).

17. For a useful but incomplete bibliography of Osgood's writings in both psychology and nuclear strategy, see Charles E. Osgood and Oliver C. S. Tzeng, *Language, Meaning, and Culture: The Selected Papers of C. E. Osgood* (New York: Praeger, 1990), 379–93.

18. Davis Howes and Charles E. Osgood, "On the Combination of Associative Probabilities in Linguistic Contexts," *American Journal of Psychology* 67 (1954): 241–58.

19. Charles E. Osgood, George J. Suci, and Percy H. Tannenbaum, *The Measurement of Meaning* (Urbana: University of Illinois Press, 1957), 20.

20. See James G. Snider and Charles E. Osgood, eds., *Semantic Differential Technique: A Sourcebook* (Chicago: Aldine, 1969).

21. Charles E. Osgood, "Reciprocal Initiatives," in *The Liberal Papers*, ed. James Roosevelt, 155–228 (Garden City, NY: Anchor Books, 1962); Elizabeth Hall, "A Conversation with Charles Osgood," *Psychology Today*, November 1973, 54.

22. Oliver C. S. Tzeng, "The Three Magnificent Themes of a Dinosaur Caper," in Osgood and Tzeng, *Language, Meaning, and Culture*, 23.

23. Charles E. Osgood, *An Alternative to War or Surrender* (1962; Urbana: University of Illinois Press, 1970), 87 (original emphasis; ellipses added).

24. Ibid., 54.

25. Ibid., 17. This argument bears strong affinities with contemporary pop-ethological works such as Robert Ardrey, *African Genesis: A Personal Investigation into the Animal Origins and Nature of Man* (New York: Atheneum, 1961).

26. Charles E. Osgood, *Perspective in Foreign Policy*, 2nd ed. (Palo Alto, CA: Pacific Books, 1966), 23 (original emphasis).

27. Osgood, *An Alternative to War or Surrender*, 56.

28. Charles E. Osgood, "Questioning Some Unquestioned Assumptions about National Security," *Social Problems* 11 (1963): 6.

29. Osgood, *Perspective in Foreign Policy*, 36. For further invocations of "rationality" in Osgood, see e.g., his *An Alternative to War or Surrender*, 58–59; "Putting the Arms Race in Reverse," *The Christian Century* 79 (1962): 566; and "Reversing the Arms Race," *Progressive*, May 1962, 27–31.

30. Osgood, *Perspective in Foreign Policy*, 26. Osgood argued that an "intermittent reinforcement" strategy might even produce *better* results than strict reinforcement, a conclusion drawn from learning theory (Osgood, *An Alternative to War or Surrender*, 163).

31. Charles E. Osgood, "Disarmament Demands GRIT," in *Toward Nuclear Disarmament and Global Security: A Search for Alternatives*, ed. Burns H. Weston, 337–44 (Boulder, CO: Westview Press, 1984), 339–42.

32. Osgood, "Suggestions for Winning the Real War with Communism," 303.

33. Ibid., 314–15. The original source for mirror imaging is the anecdotal account by social psychologist Urie Bronfenbrenner, "The Mirror Image in Soviet-American Relations: A Social Psychologist's Report," *Journal of Social Issues* 17 (1961): 45–56.

34. Marc Pilisuk and Paul Skolnick, "Inducing Trust: A Test of the Osgood Proposal," *Journal of Personality and Social Psychology* 8 (1968): 121–22.

35. Ibid., 125.

36. Amitai Etzioni, "The Kennedy Experiment," *Western Political Quarterly* 20 (1967): 361–80. This article is repeatedly cited in the GRIT literature as confirmation, despite Etzioni's caution that "it is impossible to tell, without re-running history for 'control' purposes, whether multilateral negotiations could have been successfully undertaken without the 'atmosphere' first having been improved by unilateral steps" (372).

37. Hall, "A Conversation with Charles Osgood," 54, 56.

38. Charles E. Osgood, "GRIT: A Strategy for Survival in Mankind's Nuclear Age?," in *New Directions in Disarmament*, ed. William Epstein and Bernard T. Feld, 164–72 (New York: Praeger, 1981), 171.

39. See, respectively, P. Terrence Hopmann and Timothy King, "Interactions and Perceptions in the Test Ban Negotiations," *International Studies Quarterly* 20 (1976): 105–42; Walter C. Clemens, Jr., "GRIT at Panmunjom: Conflict and Cooperation in a Divided Korea," *Asian Survey* 13 (1973): 531–59; Lawrence Juda, "Negotiating a Treaty on Environmental Modification Warfare: The Convention on Environmental Warfare and Its Impact upon Arms Control Negotiations," *International Organization* 32 (1978): 975–91; and Tony Armstrong, *Breaking the Ice: Rapprochement between East and West Germany, the United States and China, and Israel and Egypt* (Washington DC: United States Institute of Peace Press, 1993).

40. E.g., Kimberly Marten Zisk, "Soviet Academic Theories on International Conflict and Negotiations," *Journal of Conflict Resolution* 34 (1990): 684.

41. Charles E. Osgood, "The Psychologist in International Affairs," *American Psychologist* 19 (1964): 111.

42. Osgood, "GRIT," 164.

43. Charles E. Osgood, "Graduated Unilateral Initiatives for Peace," in *Preventing World War III: Some Proposals*, ed. Quincy Wright, William M. Evan, and Morton Deutsch, 161–77 (New York: Simon and Schuster, 1962), 169.

44. Charles E. Osgood, "Statement on Psychological Aspects of International Relations," in *Psychological Dimensions of Social Interaction: Readings and Perspectives*, ed. D. E. Linder, 277–85 (Reading, MA: Addison-Wesley, 1973), 278–79.

45. Quoted in James G. Blight and David A. Welch, *On the Brink: Americans and Soviets Reexamine the Cuban Missile Crisis* (New York: Hill and Wang, 1989), 198.

46. E.g., Elie Abel, *The Missile Crisis* (Philadelphia, PA: J. B. Lippincott, 1966).

47. Kennedy, *Thirteen Days*, 9.

48. Ibid., 22.

49. On the tapes, see Sheldon M. Stern, *The Week the World Stood Still: Inside the Secret Cuban Missile Crisis* (Stanford, CA: Stanford University Press, 2005), 37–38; and David A. Welch and James G. Blight, "The Eleventh Hour of the Cuban Missile Crisis: An Introduction to the ExComm Transcripts," *International Security* 12 (Winter 1987–1988): 22.

50. This conclusion flew in the face of conventional expert opinion of the efficacy of group reasoning. Paul 't Hart, *Groupthink in Government: A Study of Small Groups and Policy Failure* (Baltimore, MD: Johns Hopkins University Press, 1990), 6.

51. Irving L. Janis, *Groupthink: Psychological Studies of Policy Decisions and Fiascoes*, 2nd rev. ed. (Boston: Houghton Mifflin, 1983).

52. Irving L. Janis, *Victims of Groupthink: A Psychological Study of Foreign-Policy Decisions and Fiascoes* (Boston: Houghton Mifflin, 1972), 10. For a concise summary of the theory, see Paul 't Hart, "Irving L. Janis' Victims of Groupthink," *Political Psychology* 12 (1991): 247–78.

53. Janis, *Victims of Groupthink*, 9.

54. Kennedy, *Thirteen Days*, 11.

55. Janis, *Victims of Groupthink*, 165.

56. Gregory Moorhead, "Groupthink: Hypothesis in Need of Testing," *Group & Organization Studies* 7 (1982): 429–44.

57. Mark Schafer and Scott Crichlow, "Antecedents of Groupthink: A Quantitative

Study," *Journal of Conflict Resolution* 40 (1996): 415–35; Andrew K. Semmel and Dean Minix, "Small-Group Dynamics and Foreign Policy Decision-Making: An Experimental Approach," in *Psychological Models in International Politics*, ed. Lawrence S. Falkowski, 251–87 (Boulder, CO: Westview Press, 1979); Gregory Moorhead and John R. Montanari, "An Empirical Investigation of the Groupthink Phenomenon," *Human Relations* 39 (1986): 399–410; and Matie L. Flowers, "A Laboratory Test of Some Implications of Janis's Groupthink Hypothesis," *Journal of Personality and Social Psychology* 35 (1977): 888–96.

58. Sally Riggs Fuller and Ramon J. Aldag, "Challenging the Mindguards: Moving Small Group Analysis beyond Groupthink," in *Beyond Groupthink: Political Group Dynamics and Foreign Policy-Making*, ed. Paul 't Hart, Eric K. Stern, and Bengt Sundelius, 55–93 (Ann Arbor: University of Michigan Press, 1997), see esp. the notes.

59. Paul 't Hart, "From Analysis to Reform of Policy-Making Groups," in Hart et al., *Beyond Groupthink*, 324.

60. Paul 't Hart, Eric K. Stern, and Bengt Sundelius, "Foreign Policy-Making at the Top: Political Group Dynamics," in Hart et al., *Beyond Groupthink*, 11.

61. Charles E. Osgood and Percy H. Tannenbaum, "The Principle of Congruity in the Prediction of Attitude Change," *Psychological Review* 62 (1955): 42–55.

62. Leon Festinger, *A Theory of Cognitive Dissonance* (Stanford, CA: Stanford University Press, 1957), citation of Osgood on 8. For an explicit comparison of Festinger and Osgood, see Jack W. Brehm and Arthur R. Cohen, *Explorations in Cognitive Dissonance* (New York: John Wiley & Sons, 1962), 227–31.

63. Leon Festinger, Henry W. Riecken, and Stanley Schachter, *When Prophecy Fails: A Social and Psychological Study of a Modern Group that Predicted the Destruction of the World* (New York: Harper Torchbooks, 1956); Douglas H. Lawrence and Leon Festinger, *Deterrents and Reinforcement: The Psychology of Insufficient Reward* (Stanford, CA: Stanford University Press, 1962).

64. F. Kenneth Berrein, "Shelter Owners, Dissonance and the Arms Race," *Social Problems* 11 (1963): 87–91; Jack L. Snyder, "Rationality at the Brink: The Role of Cognitive Processes in Failures of Deterrence," *World Politics* 30 (1978): 345–65 (Cuban Missile Crisis on 365).

65. Jonathan Mercer, "Rationality and Psychology in International Politics," *International Organization* 59 (2005): 78.

CHAPTER FOUR

1. Eugene Sledge's classic World War II memoir describes the "fear and filth" of battles fought on the coral atoll of Pelau in *With the Old Breed at Peleliu and Okinawa* (London: Ebury Press, 2010). The 1968 film *Hell in the Pacific* starring Lee Marvin and Toshiro Mifune was filmed on Palau. See also Bill Sloan, *Brotherhood of Heroes: The Marines at Peleliu, 1944—The Bloodiest Battle of the Pacific War* (New York: Simon and Schuster, 2005).

2. A letter from Chester W. Nimitz to the head of the CIMA program attests to the navy's interest in scientific progress, in addition to strategic and administrative information that might be of value. "The Navy has always taken an active interest in all scientific research," he averred (letter from Admiral C.W. Nimitz to Dr. Ross Harrison [date not clear; c. late 1946/early 1947], Chairman of National Research Council, NAS-NRC Archives: ADM: Ex.Bd.: Pacific Science Board: CIMA).

3. Geopolitically, too, they had become key landmasses for US strategy. They gave the United States, in their conquest, "virtual control of the whole vast semi-circle of the

open Pacific Ocean north of the equator (the only exception [being] a small area in the Gilberts, under British sovereignty)," as a contemporary anthropologist and administrator put it, adding, "The strategic significance of being able to supervise the harbors, sea routes and airways in this Pacific zone is obvious." Therefore they earned a singular status: as a United Nations mandate of 1947 stipulated, the nascent Trust Territory was by definition and by decree "strategic." The Micronesian Trust Territory was the only "strategic trust" in existence. United Nations Trusteeship Agreement for the Former Japanese Mandated Islands, Approved by the Security Council on April 2, 1947.

4. On May 13, 1947, a naval press release announced the debut of the coordinated investigation, the most complete study ever attempted in the history of anthropology. Organizers called it "the largest cooperative research enterprise in the history of anthropology," although the participants ranged throughout the behavioral sciences, and even included a botanist. Bulletin re: CIMA Project, May 13, 1947, NAS-NRC Archives: ADM: EX Bd.: Pacific Science Board: CIMA. Originally, the project was to involve "all the sciences" but its initiators changed the focus to "primarily the geographical and human sciences (including public health)" because this would aid in problems of "practical administration" (letter from Murdock to Dr. Walter Miles, February 5, 1946, Div. of Anthropology and Psychology of NRC, NAS-NRC Archives: Div. A&P: DNRC: A&P: Com. On Anthropology of Oceania: General: 1942–43).

5. Alice Joseph, MD, and Veronica Murray, MD, *Chamorros and Carolinians of Saipan: Personality Studies* (Cambridge, MA: Harvard University Press, 1951), vii–viii.

6. The coordinated investigation's psychological test results were published as part of a raw data collecting project funded by the National Research Council in the mid-1950s and made available as microcards around the world (Bert Kaplan, ed., *Microcard Publications of Primary Records in Culture and Personality*, vol. 1 [Madison, WI: Microcard Foundation, 1956]). The initial volume contained twenty-five sets of dream-related materials contributed by seventeen field workers, including twelve sets of Rorschach tests; seven sets of thematic apperception or modified thematic apperception tests; six sets of life histories; two sets of draw-a-person, and two sentence completion tests taken from an array of nonliterate peoples.

7. It was not only Micronesia that constituted a de facto lab according to social scientific lights. A few of the many other examples of large-scale anthropological "field laboratories" operationalized in these years included a Northern California mid-sized city studied as "a social science field laboratory" by the UC Berkeley anthropology department (c. 1940s); the Peruvian village of Vicos dubbed an experimental hacienda by Cornell University (c. 1953–1961); and a spot in the New Mexico desert where five cultures (three American Indian, one Mormon, and one Spanish American) coexisted in what was described variously as an "ideal laboratory" and a "field laboratory" (c. 1949–1955). A later iteration was the "typically grandiose Kennedy-era operation called the Committee for the Comparative Study of New Nations" that focused on field sites in five 'developing' nations as they modernized. (The phrase is from Benedict Anderson, "Djojo on the Corner," *London Review of Books*, August 24, 1995, p. 19; although note that the Comparative Study originated in the Eisenhower era.) Intensive testing using the latest in social scientific methodologies freely went forward in all these field areas. On Berkeley's field lab, see William Henderson and B. W. Aginsky, "A Social Science Field Laboratory," *American Sociological Review* 6 (1941): 41–44. On Vicos, see Allan Holmberg, John Kennedy, Harold Lasswell, and Charles Lindblom, "Experimental Research in the Behavioral

Sciences and Regional Development," April 29, 1955, box 1, folder 2, Cornell-Peru Project Vicos Collection, Carl A. Kroch Library, Cornell University, Division of Rare and Manuscript Collections, Ithaca, New York. On Five Cultures as lab, see Clyde Kluckhohn, "A Comparative Study of Values in Five Cultures," in Evon Vogt, "Navaho Veterans," *Papers of the Peabody Museum of Harvard University* 41, no. 1 (1951), and Willow Roberts Powers, "The Harvard Study of Values: Mirror for Postwar Anthropology," *Journal of the History of the Behavioral Sciences* 36, no. 1 (2000): 15–29.

8. Recollection by Brewster Smith, "The American Soldier and Its Critics: What Survives the Attack on Positivism?," *Social Psychology Quarterly* 47, no. 2 (1984): 192–98, 195.

9. On the growth of methodological writing see Jennifer Platt, introduction to *A History of Sociological Research Methods in America* (Cambridge: Cambridge University Press, 1999), esp. 11–66. Platt describes methodological writing as a unique genre (an independent intellectual product) that developed in the interwar period in US sociological circles and evolved rapidly after World War II. Pre- and post-1940 is the rubicon: before 1940, methodological writing tended to be focused on how to collect data and closely examined the use of particular methods such as surveys and interviews. After 1940, methodological writing became more ambitious: "Postwar methodological writing shows much more concern with design and analysis issues" (50 n44). It tended to focus on experiment as the ideal of all research (26).

10. Style of thought (*Denkstil*) was defined by Fleck as "directed perceiving." For Fleck, not all of a thought style can be fully articulated; just so, the "situation" was widely used, but not always self-consciously attended to in methodological statements (Ludwik Fleck, *Genesis and Development of a Scientific Fact*, trans. F. Bradley and T. Trenn [1935; Chicago: University of Chicago Press, 1979], 130). On "style of thought," see also Ian Hacking, "Inaugural Lecture for the Chair of Philosophy at the Collège de France," *Economy and Society* 31, no. 1 (2001): 1–14.

11. In Dewey's words, "The function of reflective thought is, therefore, to transform a situation in which there is experienced obscurity, doubt, conflict, disturbance of some sort, into a situation that is clear, coherent, settled, harmonious" (quoted in Molly Cochran, introduction to *The Cambridge Companion to Dewey* ed. Molly Cochran [Cambridge: Cambridge, University Press, 2010], 6). On Dewey's situation, see Robert N. Grunewald, "Dewey's 'Situation' and the Ames Demonstrations," *Educational Theory* 15, no. 4 (1965): 293–304. Scholars agree that although Dewey made frequent use of the "situation," he did not systematically define it. Likewise, although he saw it as an experimental site, "he did not himself devise experimental support for 'situation.'" (293).

12. For Lewin, any observation had to be considered in the context of the "total situation," or "field," within which it was embedded. See Kurt Danziger, "Making Social Psychology Experimental: A Conceptual History, 1920–1970," *Journal of the History of the Behavioral Sciences*, 36, no. 4 (2000):340. The relevant text is Kurt Lewin, "Field Theory and Experiment in Social Psychology," in *Field Theory in Social Science* (London: Tavistock, 1939; 1952), 130–54. More generally on the rise of experimentation, see the above-mentioned special issue of the JHBS. Of particular importance was the rise of experimentation as the accepted ideal in fields such as social psychology, anthropology, and political science. On experimentation and technophilia as imperatives in modern psychology in particular, see James Capshew, "Psychologists on Site: A Reconnaissance of the Historiography of the Laboratory," *American Psychologist* 47, no. 2 (1992): 132–42; and Clare MacMartin and An-

drew Winston, "The Rhetoric of Experimental Social Psychology, 1930-1960: From Caution to Enthusiasm," *Journal of the History of the Behavioral Sciences* 36, no. 4 (2000): 349-64, 349.

13. Elliot Aronson, "Leon Festinger and the Art of Audacity," *Psychological Science* 2, no. 4 (1991): 216.

14. Matthias Jung, "John Dewey and Action," in Cochran, *The Cambridge Companion to Dewey*, 154.

15. Paul Lazarsfeld and Morris Rosenberg, *The Language of Social Research* (New York: Free Press, 1955) was the first substantial reader on methods. For an early discussion of how a social research technique can be used within a "situation," see Paul Lazarsfeld, "The Use of Panels in Social Research," *Proceedings of the American Philosophical Society* 92 (1948): 405-10.

16. Robert K. Merton, Marjorie Fiske Lowenthal, and Alberta Curtis, *Mass Persuasion: The Social Psychology of a War Bond Drive* (New York: Harper and Brothers, 1946), 3, 5.

17. Allan Holmberg, "Experimental Intervention in the Field," in *Peasants, Power, and Applied Social Change: Vicos as a Model*, ed. Henry F. Dobyns, Paul L. Doughty, and Harold Lasswell, 33-64 (London: Sage, 1971), 33.

18. Quoted in Anna McCarthy, "'Stanley Milgram, Allen Funt, and Me': Postwar Social Science and the 'First Wave' of Reality TV," in *Reality TV: Remaking Television*, ed. Susan Murray and Laurie Ouellette, 23-44 (New York: NYU Press, 2004), 27-31.

19. Herbert Simon, "Notes—On Rationality—October 7, 1965," Carnegie Mellon University Archives, Pittsburgh, PA, Box 18, 1957-1965.

20. Murray's 1949 statement of his hopes for social science is quoted in Christopher Lasch, "The Social Theory of the Therapeutic: Parsons and the Parsonians," *Haven in a Heartless World: The Family Besieged* (New York, Norton, 1977), 112.

21. On attempts to study human behavior much as physicists had studied atomic behavior, see Deborah Hammond, "Toward a Systems View of Democracy: The Society for General Systems Research" (PhD diss., University of California, Berkeley, 1997); and Ellen Herman, *The Romance of American Psychology: Political Culture in the Age of Experts* (Berkeley, University of California Press, 1995).

22. The Laboratory of Social Relations: Report for the Five Years 1946-1951, HUF 801.4156B, Harvard University Archives, Pusey Library, Cambridge, MA.

23. On the website devoted to Bales's applied scientific work, his biography describes him as an explorer of the situation foremost: "Dr. Bales began his research to answer the question, 'What is the nature of the situation'" http://www.symlog.com/internet/what_is_symlog/symlog_history-01a.htm (accessed January 28, 2012). See also "The Concept 'Situation' as a Sociological Tool" (unpublished master's dissertation, University of Oregon, 1941).

24. Robert Freed Bales, *Interaction Process Analysis* (New York: Addison Wesley, 1950), 1.

25. Letter from Parsons to Dean McGeorge Bundy recommending Bales's promotion, May 9, 1955, Harvard University Archives, Pusey Library, Cambridge, MA, UAV 801.2010.

26. Ibid. The field of small group research was not new to the postwar period but experienced tremendous growth in prestige and popularity during that time. See, e.g., Platt, *Sociological Research Methods*.

27. Loren Baritz, *Servants of Power: A History of the Use of Social Science in American Industry* (Middletown, CT, Wesleyan University Press, 1960). See also Richard Gillespie, *Manufacturing Knowledge: A History of the Hawthorne Experiments* (Cambridge: Cambridge University Press, 1993).

28. Harold B. Clemenko, "What Do You Know About You?," *Look*, May 9, 1950, re-printed in *Science Digest*, July 1950, 28, 78–82.

29. Howard Whitman, "How to Keep Out of Trouble," *Colliers*, September 25, 1948, 28–41.

30. Robert F. Bales, "The Strategy of Small Group Research," September 1950, article presented to meeting of the Sociological Research Society of the American Socio-logical Society, Denver Colorado, Rockefeller Archive Center, RF: R.G. 1.1, Series 200S, Box 521, Folder 4449.

31. Ibid.

32. Letter from Freed Bales to Conrad M. Arensberg, October 28, 1950, Rockefeller Ar-chive Center, RF: R.G. 1.1, Series 200S, Box 521, Folder 4449 (original emphasis).

33. Robert Freed Bales, "Some Statistical Problems in Small Group Research," *Journal of the American Statistical Association* 46 (1951): 315.

34. Typed memo by Bales, "Project on Training Interaction Observers," September 17, 1951, Rockefeller Archive Center, Tarrytown, NY, RF: R.G. 1.1, Series 200S, Box 521, Folder 4450.

35. Bales, *Interaction Process Analysis*, 6, 40.

36. See further discussion in Bales, *Interaction Process Analysis*.

37. Report to the Laboratory of Social Relations by Bales, November 1950, Harvard University Archives, Pusey Library, UAV 801.2010.

38. Prospectus, June 26, 1955, Rockefeller Archive Center, Tarrytown, NY, RF: R.G. 1.1, Series 200S, Box 521, Folder 4450.

39. Robert Freed Bales and Fred L. Strodtbeck, "Phases in Group Problem-Solving," *Journal of Abnormal and Social Psychology* 46, no. 4 (1951): 482–95.

40. Bales, "The Strategy of Small Group Research" (original emphasis).

41. Letter from Bales to The Laboratory Committee, October 6, 1949, Harvard Univer-sity Archives, Pusey Library, Cambridge, MA, UAV 801.2010.

42. Research proposal from Bales to Rockefeller Foundation, February 16, 1953, Rock-efeller Archive Center, Tarrytown, NY, RF: R.G. 1.1, Series 200S, Box 521, Folder 4449.

43. Resolution of RF to support Bales, April 1, 1953, Rockefeller Archive Center, Tarry-town, NY, RF: R.G. 1.1, Series 200S, Box 521, Folder 4449.

44. Robert Freed Bales, "Statement of Proposed Work," research proposal, March 20, 1952, Harvard University Archives, Pusey Library, Cambridge, MA, UAV 801.2010.

45. Robert Freed Bales, "Small-Group Theory and Research," in Robert K. Merton, Leon-ard Broom, Leonard Cottrell Jr., *Sociology Today: Problems and Prospects*, 293–305 (New York: Basic Books, 1959), 303.

46. Stuart Chapin, *Experimental Designs in Sociological Research* (New York: Harper and Brothers, 1947), viii, quoted in Platt, *A History of Sociological Research Methods in America*, 22.

47. Kant, *Critique of Pure Reason* [1781, 1787], A738/B766.

48. Robert Freed Bales, "Social Interaction," revised December 14, 1954, RAND Corpo-ration, contract P-587, p. 5. Note that at the height of Bales's Social Relations pe-riod, Parsons elected to have his own Carnegie Project on Theory tape-recorded and transcribed, then subjected to Bales's interaction process analysis, although the proj-ect itself did not, apparently, take place in the special room. See Joel Isaac, "Theorist at Work: Talcott Parsons and the Carnegie Project on Theory, 1949–1951," *Journal of the History of Ideas* 71 (2010): 287–311, 310.

49. Letter from RAND personnel manager to Parsons requesting security reference for

Bales, June 27, 1951, UAV 801.2010, Harvard University Archives, Pusey Library, Cambridge, MA.

50. Bales further stripped down his method while at RAND, it should be noted. He reduced all tasks (party-planning, chess-playing, group therapies) to a single "human relations" all-purpose task (Bales, "Social Interaction," 4). All quotations from this and the following paragraphs are from Bales, "Social Interaction," 10–12, unless otherwise noted.

51. Robert F. Bales, Merrill M. Flood, and A. S. Householder, "Some Interaction Models," RAND RM-953 (1952), 26–42. Bales's connection with military applications of his work did not persist very long after his RAND period, but his interest in choreographing order in the all-purpose conference room did. His central contribution was SYMLOG—an acronym for the system for multiple level observation of groups—which further systematized and eventually computerized the special-room procedure. He spent the remainder of his career homing in on prediction.

52. President Dwight D. Eisenhower answering questions at a press conference in the spring of 1958, quoted in Daniel Ellsberg, "The Theory and Practice of Blackmail," RAND Corporation Report, Santa Monica, CA, July 1968, 1–38, 1, accessed at http://www.rand.org/content/dam/rand/pubs/papers/2005/P3883.pdf.

53. Samuel Stouffer, "Some Thoughts About the Next Decade in Sociological Research," manuscript for annual sociology meeting, April 12, 1951, Harvard University Archives, Pusey Library, Cambridge, MA, UAV 801.2010.

54. Kenneth Boulding, preface to *The Image: Knowledge in Life and Society* (Ann Arbor: University of Michigan Press, 1956). See also E. E. Hagen, "Analytical Models in the Study of Social Systems," *American Journal of Sociology* 67 (1961): 144–51.

55. This may seem an audacious claim: for surely there was more diversity among the whole of social scientific research in the postwar period than unity, and it would be rash to claim unity under anything as elusive as "the situation." Yet our claim is that situation-based thinking was indeed common across fields, and especially across those fields inclined to unify under the banner of the behavioral sciences. Likewise, our claim is not that the situation is utterly unique and new to the Cold War period, but that it became salient during that time across fields. There was a thriving use of situations in earlier research in certain fields.

56. Discussing a related endeavor at Harvard's DSR, he argues that postwar social scientists faced the "struggle to make sense of the dizzying expansion of their tool kits" and therefore very often experienced what he calls *tool shock*, an inability to account for the "peculiarly reflexive nature of knowledge in the human sciences" (Joel Isaac, "Tool Shock: Technique and Epistemology in the Postwar Social Sciences," *History of Political Economy* 42 [annual supplement], 2010: 133–64, 135).

57. Harry Harlow, "The Nature of Love," *American Psychologist* 13 (1958): 673–85, 679.

CHAPTER FIVE

1. H. R. Haldeman, with Joseph DiMona, *The Ends of Power* (New York: Times Books, 1978), 83. Quoted in Scott D. Sagan and Jeremi Suri, "The Madman Nuclear Alert: Secrecy, Signaling, and Safety in October 1969," *International Security* 27, no. 4 (2003):156.

2. Sagan and Suri, "Madman Nuclear Alert," 150.

3. See especially William Poundstone, *Prisoner's Dilemma: John von Neumann, Game Theory, and the Puzzle of the Bomb* (New York: Doubleday, 1992).

4. See Russell L. Ackoff, David W. Conrath, and Nigel Howard, *A Model Study of the*

Escalation and De-Escalation of Conflict, report to the US Arms Control and Disarmament Agency under Contract ST-94, Management Science Center, University of Pennsylvania, March 1, 1967; and Robert J. Aumann, John C. Harsanyi, Michael Maschler, John P. Mayberry, Reinhard Selten, Herbert Scarf, and Richard E. Stearns, *Models of Gradual Reduction of Arms*, submitted to the Arms Control and Disarmament Agency, summary of final report on contract No. ACDA / ST-116 (Princeton, NJ: Mathematica, 1967).

5.   Cf. Anatol Rapoport and Melvin Guyer, "A Taxonomy of 2×2 Games," *General Systems* 11 (1966): 203–14.

6.   This literature focuses on exploring prisoner's dilemma games under various combinations of indefinite or infinite repetition, incomplete information, and intertemporal discount rates; see especially Robert J. Aumann and Lloyd S. Shapley, "Long-Term Competition—A Game-Theoretic Analysis," in *Essays in Game Theory In Honor of Michael Maschler*, ed. Nimrod Megiddo (New York: Springer-Verlag, 1994), 1–27; Ariel Rubinstein, "Equilibrium in Supergames with the Overtaking Criterion," *Journal of Economic Theory* 21, no. 1 (1979): 1–9; David M. Kreps, Paul Milgrom, John Roberts and Robert Wilson, "Rational Cooperation in the Finitely Repeated Prisoners' Dilemma," *Journal of Economic Theory* 27, no. 2 (1982): 245–52; and Dilip Abreu, "On the Theory of Infinitely Repeated Games with Discounting," *Econometrica* 56, no. 2 (1988): 383–96.

7.   Merrill M. Flood, "Some Experimental Games," RAND RM-789-1 (June 20, 1952), 17.

8.   Poundstone, *Prisoner's Dilemma*, 117.

9.   "A Two-Person Dilemma." Merrill M. Flood Papers, Box 1, Folder: "Notes, 1929–1967," Bentley Historical Library, University of Michigan, Ann Arbor, MI.

10.   For analysis of the narratives surrounding the prisoner's dilemma game, see Mary S. Morgan, "The Curious Case of the Prisoner's Dilemma: Model Situation? Exemplary Narrative?" in *Science Without Laws: Model Systems, Cases, Exemplary Experiments* ed. Angela N. H. Creager, Elizabeth Lunbeck, and M. Norton Wise, 157–85 (Durham, NC: Duke University Press, 2007).

11.   John von Neumann and Oskar Morgenstern, *Theory of Games and Economic Behavior* 2nd ed. (Princeton, NJ: Princeton University Press, 1953), 33.

12.   Some games apparently do not possess solutions in the sense of von Neumann and Morgenstern; see e.g., W. F. Lucas, "The Proof that a Game may not have a Solution," RAND RM-5543-PR (January 1968).

13.   On Dantzig's visit to Princeton and its aftermath, see e.g., George Dantzig, "Reminiscences about the Origins of Linear Programming," *Operations Research Letters* 1 (1982): 43–48; and J. K. Lenstra, A. H. G. Rinnooy Kan, and A. Schrijver, eds., *History of Mathematical Programming: A Collection of Personal Reminiscences* (Amsterdam: North-Holland, 1991).

14.   Cf. Philip Mirowski, "When Games Grow Deadly Serious: The Military Influence upon the Evolution of Game Theory," in *Economics and National Security: A History of their Interaction* (Durham, NC: Duke University Press, 1991).

15.   On the history of computing hardware at RAND, see Willis H. Ware, "RAND Contributions to the Development of Computing," http://www.rand.org/about/history/ware.html (accessed June 30, 2010).

16.   George W. Brown, "History of RAND's Random Digits—Summary" RAND P-113 (June 1949), 5. On random number production and its application to the solution of differential equations and the simulation of random processes at RAND and

elsewhere see e.g., N. Metropolis, "The Beginning of the Monte Carlo Method" *Los Alamos Science* Special Issue (1987): 125–29; Peter Galison, *Image and Logic: A Material Culture of Microphysics* (Chicago: University of Chicago Press, 1997), chapter 8; Sharon Ghamari-Tabrizi, *The Worlds of Herman Kahn: The Intuitive Science of Thermonuclear War* (Cambridge, MA: Harvard University Press, 2005), 133–36.

17. Olaf Helmer, "Recent Developments in the Mathematical Theory of Games," RAOP-16 (April 30, 1948), 16–18.

18. A. W. Tucker and R. D. Luce, eds., *Contributions to the Theory of Games*, vol. IV (Princeton, NJ: Princeton University Press, 1959), 2.

19. John F. Nash, "Non-Cooperative Games" (PhD diss., Princeton University, May 1950), 1.

20. Ibid., 3.

21. See "A Two-Person Dilemma" in ibid. (underlining in the original).

22. Flood, "Some Experimental Games," 24.

23. See, e.g., Martin Shubik, "Game Theory at Princeton, 1949–1955: A Personal Reminiscence" in *Toward a History of Game Theory*, ed. E. Roy Weintraub, 151–63 (Durham, NC: Duke University Press, 1992).

24. Flood, "Some Experimental Games," 24.

25. John von Neumann, "On the Theory of Games of Strategy" in Tucker and Luce, *Contributions to the Theory of Games*, 4:23.

26. See Anatol Rapoport, *Certainties and Doubts: A Philosophy of Life* (Toronto: Black Rose Books, 2000), chapters 8–9.

27. Anatol Rapoport and Albert M. Chammah, *Prisoner's Dilemma: A Study in Conflict and Cooperation* (Ann Arbor: University of Michigan Press, 1965), 11.

28. Ibid., vi.

29. Ibid., 6.

30. Ibid., 198–99.

31. Ibid., 199.

32. Erica Frydenberg, *Morton Deutsch: A Life and Legacy of Mediation and Conflict Resolution* (Brisbane: Australian Academic Press, 2005), 56.

33. Kurt Lewin and Ronald Lippitt, "An Experimental Approach to the Study of Autocracy and Democracy: A Preliminary Note" *Sociometry* 1.3/4 (January–April 1938), 292–300; see also Kurt Lewin, Ronald Lippitt, and Ralph K. White, "Patterns of Aggressive Behavior in Experimentally Created 'Social Climates'" *Journal of Social Psychology* 10 (1939): 271–99.

34. On Lewin, see Marvin Ross Weisbord, *Productive Workplaces Revisited: Dignity, Meaning, and Community in the 21st Century* (San Francisco: Jossey-Bass, 2004), chapters 4–5; Alfred J. Marrow, *The Practical Theorist: The Life and World of Kurt Lewin* (New York: Basic Books, 1969); William Graebner, *The Engineering of Consent: Democracy and Authority in Twentieth-Century America* (Madison: University of Wisconsin Press, 1987).

35. Frydenberg, *Morton Deutsch*, 58. See also Morton Deutsch and Mary Evans Collins, *Interracial Housing: A Psychological Evaluation of a Social Experiment* (Minneapolis: University of Minnesota Press, 1951).

36. See M. Deutsch, *Conditions Affecting Cooperation* (final technical report for the Office of Naval Research, Contract NONR-285[10], February 1957), cited in Morton Deutsch, "Trust and Suspicion," *Journal of Conflict Resolution* 2, no. 4 (December 1958), 265–79.

37. Deutsch, "Trust and Suspicion," 269–70.

38. Ibid., 270.

39. Frydenberg, *Morton Deutsch*, 67.

40. See Anatol Rapoport, "Lewis F. Richardson's Mathematical Theory of War," *Conflict Resolution* 1, no. 3 (September 1957): 249–99.

41. Rapoport, "Lewis F. Richardson's Mathematical Theory of War," 284–85.

42. Rapoport, *Certainties and Doubts*, 113.

43. See, e.g., Anatol Rapoport, *Strategy and Conscience* (New York: Harper and Row, 1964).

44. See, e.g., Paul Crook, *Darwinism, War, and History: the Debate over the Biology of War from the "Origin of Species" to the First World War* (Cambridge: Cambridge University Press, 1994); Gregg Mitman, *The State of Nature: Ecology, Community, and American Social Thought, 1900–1950* (Chicago: University of Chicago Press, 1992).

45. See, e.g., the review paper by Sir Julian Huxley, "Introduction: A Discussion of Ritualization of Behavior in Animals and Man," *Philosophical Transactions of the Royal Society of London Series B (Biological Sciences)* 251, no. 772 (1966): 249–71.

46. On this shift within biology and genetics, see, e.g., Daniel J. Kevles, *In the Name of Eugenics: Genetics and the Uses of Human Heredity* (New York: Knopf, 1985).

47. Robert Ardrey, *African Genesis: A Personal Investigation into the Animal Origins and Nature of Man* (New York: Atheneum, 1961); see also Robert Ardrey, *The Territorial Imperative: A Personal Inquiry into the Animal Origins of Property and Nations* (New York: Atheneum, 1966); and Desmond Morris, *The Naked Ape: A Zoologist's Study of the Human Animal* (New York: McGraw-Hill, 1967).

48. See, e.g., Lily E. Kay, *Who Wrote the Book of Life? A History of the Genetic Code* (Stanford, CA: Stanford University Press, 2000); Evelyn Fox Keller, *The Century of the Gene* (Cambridge, MA: Harvard University Press, 2000).

49. W. D. Hamilton, "The Genetical Evolution of Social Behavior, I," *Journal of Theoretical Biology* 7 (1964): 1–26; Hamilton, "The Genetical Evolution of Social Behavior, II," *Journal of Theoretical Biology* 7 (1964): 27–52.

50. W. D. Hamilton, "Extraordinary Sex Ratios," *Science* (New Series) 156, no. 3774 (April 28, 1967): 477–88.

51. Hamilton to Price, March 21, 1968, Item KPX1_5.5 Price Papers, Hamilton Archive, British Library, London (hereafter referred to as "Price Papers").

52. Anatol Rapoport, "Escape from Paradox" *Scientific American*, July 1967, 50–56.

53. Hamilton to Price, March 21, 1968, Item KPX1_5.5, Price Papers.

54. See, e.g., ibid.

55. W. D. Hamilton, "Selection of Selfish and Altruistic Behavior in some Extreme Models," in *Man and Beast: Comparative Social Behavior*, ed. J. F. Eisenberg and Wilton S. Dillon, 59–91 (Washington DC: Smithsonian Institution Press, 1971).

56. Hamilton, "Selection of Selfish and Altruistic Behavior," 82–83.

57. Hamilton to Wilton S. Dillon, Smithsonian Institution, January 30, 1970, Hamilton Papers (no number yet assigned), British Library, London, UK.

58. Robert Trivers, "The Evolution of Reciprocal Altruism," *Quarterly Review of Biology* 46 (1971): 35–57.

59. On the prisoner's dilemma tournaments see Robert Axelrod, "Effective Choice in the Prisoner's Dilemma," *Journal of Conflict Resolution* 24 (1980): 3–25; Axelrod, "More Effective Choice in the Prisoner's Dilemma," *Journal of Conflict Resolution* 24 (1980): 379–403.

60. Robert Axelrod and William D. Hamilton, "The Evolution of Cooperation," *Science* 211 (1981):1390–96; and Robert Axelrod, *The Evolution of Cooperation* (New York: Basic Books, 1984).

CHAPTER SIX

1. For the following, see Leopold Labedz, *Poland under Jaruzelski: A Comprehensive Sourcebook on Poland During and After Martial Law* (New York: Scribner, 1984); George Sanford, *Military Rule in Poland: The Rebuilding of Communist Power, 1981– 1983* (London: Croom Heim, 1986).

2. Amos Tversky and Daniel Kahneman, "Extensional versus Intuitive Reasoning: The Conjunction Fallacy in Probability Judgment," *Psychological Review* 90 (1983): 308.

3. See, e.g., Mike Oaksford and Nick Chater, "Human Rationality and the Psychology of Reasoning: Where Do We Go from Here?" *British Journal of Psychology* 92 (2001): 193–216; Alan R. Anderson and Nuel D. Belnap, *Entailment: The Logic of Relevance and Necessity*, vol. 1 (Princeton, NJ: Princeton University Press, 1975); Bruno de Finetti, "La prévision: Ses lois logiques, ses sources subjectives," *Annales de l'Institute Henri Poincaré* 7 (1937): 1–68; Frank P. Ramsey, *The Foundations of Mathematics and Other Logical Essays* (London: Routledge & Kegan Paul, 1931); John von Neumann and Oskar Morgenstern, *Theory of Games and Economic Behavior* (Princeton, NJ: Princeton University Press, 1944); Leonard J. Savage, *Foundations of Statistics* (New York: Wiley, 1954).

4. Cited in "Peter Wason: Obituary," *Daily Telegraph*, April 21, 2003.

5. Philip Johnson-Laird, "Peter Wason: Obituary," *Guardian*, April 25, 2003.

6. See Keith Stenning and Michiel van Lambalgen, "The Natural History of Hypotheses about the Selection Task," in *Psychology of Reasoning*, ed. Ken Manktelow and Man C. Chung, 127–56 (Hove & New York: Psychology Press, 2004).

7. Floris Heukelom, "Kahneman and Tversky and the Making of Behavioral Economics" (PhD diss., University of Amsterdam, 2009).

8. Savage, *Foundations of Statistics*; Ward Edwards, "Behavioral Decision Theory," *Annual Review of Psychology* 12 (1961): 473–98.

9. Daniel Kahneman, "Autobiography," http://nobelprize.org/nobel_prizes/econom ics/laureates/2002/kahneman-autobio.html (accessed July 26, 2012).

10. Daniel Kahneman and Amos Tversky, "On the Psychology of Prediction," *Psychological Review* 80 (1973): 237–51.

11. See Daniel Kahneman, "Maps of Bounded Rationality: Psychology for Behavioral Economics," *American Economic Review* 93 (2003): 1449–75; Daniel Kahneman and Amos Tversky, "Prospect Theory: An Analysis of Decision under Risk," *Econometria* 47 (1979): 263–92.

12. Bärbel Inhelder and Jean Piaget, *The Growth of Logical Thinking from Childhood to Adolescence* (New York: Basic Books, 1958), 305 (translation modified). The English-language edition renders the passage as "reasoning is nothing but the propositional calculus," but the original says, "le raisonnement n'est que le calcul comme tel que comportent les operations propositionelles" (Bärbel Inhelder and Jean Piaget, *De la logique de l'enfant à la logique de l'adolescence* [Paris: PUF, 1955], 270).

13. Peter C. Wason, "Reasoning about a Rule," *Quarterly Journal of Experimental Psychology* 20 (1968): 273–81.

14. Peter C. Wason, "Reasoning," in *New Horizons in Psychology*, ed. Brian M. Foss, 135–51 (Harmondsworth: Penguin, 1966), 146.

15. Ibid., 274.

16. Ibid., 147.

17. Karl R. Popper, *The Logic of Scientific Discovery* (New York: Basic Books, 1959).

18. E.g., Richard E. Nisbett and Lee Ross, *Human Inference: Strategies and Shortcomings of Social Judgement* (Englewood Cliffs, NJ: Prentice-Hall, 1980).

19. Tversky and Kahneman, "Extensional versus Intuitive Reasoning."

20. Ibid., 299.

21. Amos Tversky and Daniel Kahneman, "On the Reality of Cognitive Illusions," *Psychological Review* 103 (1996): 582–91.

22. Ibid., 313.

23. Ward Casscells, Arno Schoenberger, and Thomas B. Grayboys, "Interpretation by Physicians of Clinical Laboratory Results," *New England Journal of Medicine* 299 (1978): 999–1001.

24. Amos Tversky and Daniel Kahneman, "Judgment under Uncertainty: Heuristics and Biases," *Science* 185 (1974): 1124–31; Daniel Kahneman, Paul Slovic, and Amos Tversky, eds., *Judgment under Uncertainty: Heuristics and Biases* (New York: Cambridge University Press, 1982); Thomas Gilovich, Dale W. Griffin, and Daniel Kahneman, eds., *Heuristics and Biases: The Psychology of Intuitive Judgment* (Cambridge: Cambridge University Press, 2002).

25. Amos Tversky and Daniel Kahneman, "Belief in the Law of Small Numbers," *Psychological Bulletin* 2 (1971): 105–10.

26. Amos Tversky and Daniel Kahneman, "Rational Choice and the Framing of Decisions," *The Journal of Business* 59 (1986): S273.

27. Richard B. Nisbett and Eugene Borgida, "Attribution and the Psychology of Prediction," *Journal of Personal and Social Psychology* 32 (1975): 935.

28. Peter C. Wason, "Realism and Rationality in the Selection Task," in *Thinking and Reasoning*, ed. Jonathan St. B. T. Evans (London: Routledge & Kegan Paul, 1983), 59.

29. Massimo Piattelli-Palmarini, *Inevitable Illusions: How Mistakes of Reason Rule Our Minds* (New York, 1994): 132; see also Stuart Sutherland, *Irrationality* (London: Pinker & Martin, 1992).

30. Daniel Reisberg, *Cognition: Exploring the Science of the Mind* (New York: W. W. Norton, 1997), 469–70.

31. E.g., Thomas Gilovich, *How We Know What Isn't So: The Fallibility of Human Reason in Everyday Life* (New York: Free Press, 1991); Sutherland, *Irrationality*; Piatelli-Palmarini, *Inevitable Illusions*.

32. Baruch Fischhoff, Paul Slovic, Sarah Lichtenstein, Stephen Read, and Barbara Combs, "How Safe Is Safe Enough? A Psychometric Study of Attitudes towards Technological Risks and Benefits," *Policy Sciences* 9 (1978): 127–52; Nancy Kanwisher, "Cognitive Heuristics and American Security Policy," *Journal of Conflict Resolution* 33 (1989): 652–75; Mark L. Haas, "Prospect Theory and the Cuban Missile Crisis," *International Studies Quarterly* 45 (2001): 241–70; Rose McDermott, "The Psychological Ideas of Amos Tversky and Their Relevance for Political Science," *Journal of Theoretical Politics* 13 (2001): 5–33; Rose McDermott, "Arms Control and the First Reagan Administration: Belief-systems and Policy Choices," *Journal of Cold War Studies* 4 (2002): 29–59.

33. George A. Quattrone and Amos Tversky, "Self-Deception and the Voter's Illusion," *Journal of Personality and Social Psychology* 46 (1984): 719–36.

34. Philip E. Tetlock, "Theory-Driven Reasoning about Plausible Pasts and Probable Futures in World Politics," in *Heuristics and Biases*, ed. Thomas Gilovich, Dale W. Griffin, and Daniel Kahneman, 749–62 (Cambridge: Cambridge University Press, 2002).

35. Paul K. Davis and John Arquilla, "Thinking about Opponent Behavior in Crisis and Conflict: A Generic Model for Analysis and Group Discussion," RAND AD–A253 258 (July 29, 1992).

36. Ariel S. Levy and Glen Whyte, "A Cross-cultural Explanation of the Reference Dependence of Crucial Group Decision under Risk: Japan's 1941 Decision for War," *Journal of Conflict Resolution* 41 (1997): 792–813.

37. Haas, "Prospect Theory and the Cuban Missile Crisis."

38. Philip E. Tetlock, Charles B. McGuire, and Gregory Mitchell, "Psychological Perspectives on Nuclear Deterrence," *Annual Review of Psychology* 42 (1991): 239–76.

39. Rose McDermott, *Risk Taking in International Politics: Prospect Theory in Postwar American Foreign Policy* (Ann Arbor, MI: University of Michigan Press, 1998).

40. Richard Spielman, "Crisis in Poland," *Foreign Policy* 49 (1982–83), 25, 30–31.

41. Kanwisher, "Cognitive Heuristics," 652.

42. Tversky and Kahneman, "Extensional versus Intuitive Reasoning."

43. Kanwisher, "Cognitive Heuristics," 655.

44. Ellen J. Langer, *The Psychology of Control* (Beverly Hills, CA: Sage Publications, 1983).

45. Tversky and Kahneman, "Prospect Theory." A refined, "cumulative" version of the theory was presented in Amos Tversky and Daniel Kahneman, "Advances in Prospect Theory: Cumulative Representation of Uncertainty," *Journal of Risk and Uncertainty* 5 (1992): 297–323. That prospect theory tries to "repair" classical expected utility theory is expressed by Reinhard Selten, "What Is Bounded Rationality?" in *Bounded Rationality: The Adaptive Toolbox,* ed. Gerd Gigerenzer and Reinhard Selten, 13–36 (Cambridge, MA: MIT Press 2001).

46. George Quattrone and Amos Tversky, "Contrasting Psychological and Rational Analyses of Political Choice," *American Political Science Review* 82 (1988): 719–36.

47. McDermott, "Amos Tversky," 35.

48. Ibid., 58.

49. Richard H. Immerman, "Psychology," *Journal of American History* 77 (1990): 177.

50. L. Jonathan Cohen, "On the Psychology of Prediction: Whose Is the Fallacy?" *Cognition* 7 (1979): 385–407; Cohen, "Whose is the Fallacy? A Rejoinder to Kahneman and Tversky," *Cognition* 8 (1980): 89–92; Daniel Kahneman and Amos Tversky, "On the Interpretation of Intuitive Probability: A Reply to Jonathan Cohen," *Cognition* 7 (1980): 409–11.

51. L. Jonathan Cohen, "Can Human Irrationality Be Experimentally Demonstrated?" *Behavioral and Brain Sciences* 4 (1981): 317–31 (comments and responses, 331–59).

52. Cohen here applied well-known ideas of Nelson Goodman, *Fact, Fiction, and Forecast* (London: Athlone, 1954).

53. Cohen, "Human Irrationality," 318–23.

54. Reported by Nisbett and Ross, *Human Inference,* 249.

55. Helmut Jungermann, "The Two Camps on Rationality," in *Decision Making under Uncertainty,* ed. Roland W. Scholz, 63–86 (Amsterdam: North-Holland, 1983); see also Jonathan St. B. T. Evans, "Theories of Human Reasoning: The Fragmented State of the Art," *Theory & Psychology* 1 (1991): 83–105; Edward Stein, *Without Good Reason: The Rationality Debate in Philosophy and Cognitive Science* (Oxford: Oxford University Press, 1996).

56. Richard Samuels, Stephen Stich, and Michael Bishop, "Ending the Rationality Wars: How to Make Disputes about Human Rationality Disappear," in *Common Sense, Reasoning and Rationality,* ed. Renée Elio, 236–68 (Oxford: Oxford University Press, 2002).

57. Full disclosure: some of us (Lorraine Daston and Thomas Sturm) have published with one major participant in the debate, Gerd Gigerenzer. This puts us in a situa-

tion where we must be careful to avoid partiality. This is why the actual arguments in the debate about rationality are analyzed more closely than may be usual in a volume like this. We decided that avoiding a major topic would have been worse, and we have tried to present the dispute—and not to solve it—*sine ira et studio*.

58. Lance J. Rips and S. L. Marcus, "Supposition and the Analysis of Conditional Sentences," in *Cognitive Processes in Comprehension*, ed. Marcel A. Just and Patricia A. Carpenter, 185–220 (Hillsdale, NJ: Erlbaum, 1977).

59. Cameron R. Peterson and Lee R. Beach, "Man as an Intuitive Statistician," *Psychological Bulletin* 68 (1967): 29–46.

60. Peterson and Beach, "Man as an Intuitive Statistician," 42.

61. E.g., Daniel Kahneman and Amos Tversky, "Subjective Probability: A Judgment of Representativeness," *Cognitive Psychology* 3 (1972), 449–50.

62. Jay J. Christensen-Szalanski and Lee R. Beach, "The Citation Bias: Fad and Fashion in the Judgment and Decision Literature," *American Psychologist* 39 (1984): 75–78.

63. E.g., Lola L. Lopes, "Performing Competently," *Behavioral and Brain Sciences* 4 (1981): 343–44.

64. Amos Tversky and Daniel Kahneman, "Judgment under Uncertainty: Heuristics and Biases," *Science* 185 (1974): 1124–31.

65. Lola L. Lopes, "The Rhetoric of Irrationality," *Theory & Psychology* 1 (1991): 67.

66. Richard A. Griggs and James R. Cox, "The Elusive Thematic Materials Effects in Wason's Selection Task," *British Journal of Psychology* 73 (1982): 407–20.

67. Ibid.

68. Cohen, "Can Human Irrationality Be Experimentally Demonstrated?," 327–28.

69. William Kneale and Martha Kneale, *The Development of Logic* (Oxford: Clarendon Press, 1962); Gerd Gigerenzer, Zeno Swijtink, Theodore Porter, Lorraine Daston, John Beatty, and Lorenz Krüger, *The Empire of Chance* (Cambridge: Cambridge University Press, 1989).

70. Cohen, "On the Psychology of Prediction." See also L. Jonathan Cohen, *The Probable and the Provable* (Oxford: Clarendon Press, 1977); Cohen, "Bayesianism versus Baconianism in the Evaluation of Medical Diagnostics," *British Journal for the Philosophy of Science* 31 (1980): 45–62; Cohen, "Some Historical Remarks on the Baconian Concept of Probability," *Journal of the History of Ideas* 41 (1980): 219–31.

71. Michael H. Birnbaum, "Base Rates in Bayesian Inference: Signal Detection Analysis of the Cab Problem," *American Journal of Psychology* 96 (1983): 85–94.

72. For further views, see George Botterill and Peter Carruthers, *The Philosophy of Psychology* (Cambridge: Cambridge University Press, 1999), 105–30; Evans, "Theories of Human Reasoning"; Evans, "Reasoning with Bounded Rationality," *Theory & Psychology* 2 (1992): 237–42; Robert Nozick, *The Nature of Rationality* (Princeton, NJ: Princeton University Press, 1993); Stein, *Without Good Reason*.

73. Jonathan St. B. T. Evans and David E. Over, *Rationality and Reasoning* (Hove, UK: Psychology Press, 1996).

74. For an overview, see Keith E. Stanovich and Richard F. West, "Evolutionary versus Instrumental Goals: How Evolutionary Psychology Misconceives Human Rationality," in *Evolution and the Psychology of Thinking: The Debate*, ed. David E. Over (New York: Psychology Press, 2003), 183 (table 3).

75. Herbert A. Simon, *Models of Man* (New York: Wiley, 1957); Simon, *Models of Bounded Rationality* (Cambridge, MA: MIT Press, 1982).

76. George Quattrone and Amos Tversky, "Contrasting Psychological and Rational Analyses of Political Choice," *American Political Science Review* 82 (1988): 719–36;

Daniel Kahneman, "Maps of Bounded Rationality: Psychology for Behavioral Economics," *American Economic Review* 93 (2003): 1449–75; Gilovich, *How We Know What Isn't So.*

77. Gerd Gigerenzer, "Striking a Blow for Sanity in Theories of Rationality," in *Models of a Man: Essays in Memory of Herbert A. Simon,* ed. Mie Augier and James G. March, 389–409 (Cambridge, MA: MIT Press, 2004).

78. Herbert A. Simon, "Invariants of Human Behavior," *Annual Review of Psychology* 41 (1990), 7. The metaphor is already used in writings since the 1960s. See Alan Newell and Herbert Simon, "Task Environments," *Complex Information Processing,* paper 94 (1967), 4 (Herbert A. Simon Collection,Carnegie Mellon University, Pittsburgh, PA, http://diva.library.cmu.edu/webapp/simon/item.jsp?q=/box00009/fld00612/bd 10001/doc0001/, accessed July 26, 2012); Herbert A. Simon, "Rationality in Psychology and Economics" (unpublished manuscript, October 19, 1985, 21), Herbert A. Simon Collection, Carnegie Mellon University, Pittsburgh, PA http://diva.library .cmu.edu/webapp/simon/item.jsp?q=/box00057/fld04341/bd10013/doc0002/ (accessed July 26, 2012).

79. Gerd Gigerenzer, *Adaptive Thinking* (New York: Oxford University Press, 2000).

80. E.g., Robert Axelrod and William D. Hamilton, "The Evolution of Cooperation," *Science* 211 (1981): 1390–96; Robert Axelrod, *The Evolution of Cooperation* (New York: Basic Books, 1984); John Maynard Smith, *Evolution and the Theory of Games* (Cambridge: Cambridge University Press, 1982).

81. Leda Cosmides, "The Logic of Social Exchange: Has Natural Selection Shaped how Humans Reason? Studies with the Wason Selection Task," *Cognition* 31 (1989): 187–276; Leda Cosmides and John Tooby, "Cognitive Adaptations for Social Exchange," in *The Adapted Mind* ed. Jerome H. Barkow, Leda Cosmides, and John Tooby, 163–228 (Oxford: Oxford University Press, 1992).

82. Kanwisher, "Cognitive Heuristics."

83. Next to the objections mentioned, see also Klaus Fiedler, "The Dependence of the Conjunction Fallacy on Subtle Linguistic Factors," *Psychological Research* 50 (1988): 123–29.

84. Tversky and Kahneman, "Extensional versus Intuitive Reasoning," 308.

85. Tetlock has noted that political experts, even if encouraged to make unequivocal forecasts, tend to make only conditional ones, and complex ones at that: "If X1 and X2, then Y1; if X3, X4 and X5, then Y2." See Philip E. Tetlock, "Good Judgment in International Politics: Three Psychological Perspectives," *Political Psychology* 13 (1992): 529.

86. McDermott, "Arms Control and the First Reagan Administration."

87. Peter Suedfeld, "Cognitive Managers and Their Critics," *Political Psychology* 13 (1992): 441.

88. Suedfeld, "Cognitive Managers," 435.

89. E.g., Tetlock, "Good Judgment in International Politics"; Tetlock, *Expert Political Judgment: How Good Is It? How Can We Know?* (Princeton, NJ: Princeton University Press, 2005).

90. Tetlock, "Good Judgment in International Politics," 523–28.

91. Ibid., 527.

92. Ibid., 528.

93. Philip E. Tetlock, "Correspondence and Coherence: Indicators of Good Judgment in World Politics," in *Thinking: Psychological Perspectives on Reasoning,* ed. David Hardman and Laura Macchi, 233–50 (Chichester: Wiley, 2003), 233.

94. Andrei Amalrik, *Will the Soviet Union Survive until 1984?* (New York: Harper & Row, 1970).

95. John Lewis Gaddis, "International Relations Theory and the End of the Cold War," *International Security* 17 (1992–93), 5–6.

EPILOGUE

1. Julius Margolis, "Discussion," in *Strategic Interaction and Conflict: Original Papers and Discussion*, ed. Kathleen Archibald (Berkeley: Institute of International Studies, 1966), 137.

2. Thomas Schelling, "Discussion," in Archibald, *Strategic Interaction*, 150.

3. One of the ur-texts of this move came from anthropology, a lesser participant in Cold War rationality: Clifford Geertz, *The Interpretation of Cultures: Selected Essays* (New York: Basic Books, 1973).

4. The program derived its name from the influential Daniel Kahneman, Paul Slovic, and Amos Tversky, eds., *Judgment under Uncertainty: Heuristics and Biases* (Cambridge: Cambridge University Press, 1982).

5. For further examples of methodological mania in the human sciences during this period, see Jamie Cohen-Cole, "Instituting the Science of the Mind: Intellectual Economies and Disciplinary Exchange at Harvard's Center for Cognitive Studies," *British Journal for the History of Science* 40 (2007): 567–97.

6. Alfred R. Mele and Piers Rawling, eds., *The Oxford Handbook of Rationality* (Oxford: Oxford University Press, 2004).

BIBLIOGRAPHY

Abella, Alex. *Soldiers of Reason: The Rand Corporation and the Rise of American Empire*. Orlando, FL: Harcourt, 2008.

Abreu, Dilip. "On the Theory of Infinitely Repeated Games with Discounting." *Econometrica* 56, no. 2 (1988): 383–96.

Ackoff, Russell L., David W. Conrath, and Nigel Howard. *A Model Study of the Escalation and De-Escalation of Conflict*. Report to the U.S. Arms Control and Disarmament Agency under Contract ST-94, Management Science Center, University of Pennsylvania, March 1, 1967.

Ainsworth, Mary, and Barbara Wittig. "Attachment and Exploratory Behavior of One-Year-Olds." In *Determinants of Infant Behaviour* 4, ed. B. M. Foss. London: Methuen, 1969, 111–23.

Akera, Atsushi. *Calculating a Natural World: Scientists, Engineers, and Computers during the Rise of U.S. Cold War Research*. Cambridge, MA: MIT Press, 2007.

Albers, Donald, and Constance Reid. "An Interview with George B. Dantzig: The Father of Linear Programming." *College Mathematics Journal* 17, no. 4 (1986): 293–314.

Allison, Graham T. *Essence of Decision: Explaining the Cuban Missile Crisis*. Boston: Little, Brown & Company, 1971.

Amadae, Sonia M. *Rationalizing Capitalist Democracy: The Cold War Origins of Rational Choice Liberalism*. Chicago: University of Chicago Press, 2003.

Amalrik, Andrei. *Will the Soviet Union Survive until 1984?* New York: Harper & Row, 1970.

Anderson, Alan R., and Nuel D. Belnap. *Entailment: The Logic of Relevance and Necessity*. Vol. 1. Princeton, NJ: Princeton University Press, 1975.

Anderson, Benedict. "Djojo on the Corner." *London Review of Books*, August 24, 1995, 19–20.

Antognazza, Maria Rosa. *Leibniz: An Intellectual Biography*. Cambridge: Cambridge University Press, 2009.

Arago, François, and Arthur Condorcet-O'Connor, eds. *Oeuvres de Condorcet*, 12 vols. Paris: Firmin Didot Frères, 1847–1849.

Archibald, Kathleen. Introduction to *Strategic Interaction and Conflict*, edited by Kathleen Archibald, v–viii. Berkeley: University of California Institute of International Studies, 1966.

Ardrey, Robert. *African Genesis: A Personal Investigation into the Animal Origins and Nature of Man*. New York: Atheneum, 1961.

——. *The Territorial Imperative: A Personal Inquiry into the Animal Origins of Property and Nations*. New York: Atheneum, 1966.

Armstrong, Tony. *Breaking the Ice: Rapprochement between East and West Germany, the United States and China, and Israel and Egypt*. Washington DC: United States Institute of Peace Press, 1993.

Arrow, Kenneth J. *Social Choice and Individual Values*. New Haven, CT: Yale University Press, [1953] 1963.

Augier, Mie, and James March, eds. *Models of a Man*: Essays in Memory of Herbert A. Simon. Cambridge, MA: MIT Press, 2004.

Aumann, Robert J., John C. Harsanyi, Michael Maschler, John P. Mayberry, Reinhard Selten, Herbert Scarf, and Richard E. Stearns. *Models of Gradual Reduction of Arms*. Summary of Final Report on Contract No. ACDA / ST-116 submitted to the Arms Control and Disarmament Agency. Princeton, NJ: Mathematica, September 1967.

Aumann, Robert J., and Lloyd S. Shapley. "Long-Term Competition—A Game-Theoretic Analysis." In *Essays in Game Theory in Honor of Michael Maschler*, edited by Nimrod Megiddo. New York: Springer-Verlag, 1994.

Axelrod, Robert. "Effective Choice in the Prisoner's Dilemma." *Journal of Conflict Resolution* 24 (1980): 3–25.

——. *The Evolution of Cooperation*. New York: Basic Books, 1984.

——. "More Effective Choice in the Prisoner's Dilemma." *Journal of Conflict Resolution* 24 (1980): 379–403.

Axelrod, Robert, and William D. Hamilton. "The Evolution of Cooperation." *Science* 211 (1981): 1390–96.

Babbage, Charles. *On the Economy of Machinery and Manufactures*. London: Charles Knight, 1835.

——. *The Ninth Bridgewater Treatise. A Fragment*. London: John Murray, 1837.

——. *Passages from the Life of a Philosopher* [1864], excerpts reprinted in *Charles Babbage and His Calculating Engines*, edited by Philip Morrisson and Emily Morrison, 5–157. New York: Dover, 1961.

Baker, Keith Michael. *Condorcet: From Natural Philosophy to Social Mathematics*. Chicago: University of Chicago Press, 1975.

——. "An Unpublished Essay by Condorcet on Technical Methods of Classification." *Annals of Science* 18 (1962): 99–123.

Baldwin, Robert Freed. *Rules and Government*. Oxford: Clarendon Press, 1995.

Bales, Robert Freed. *Interaction Process Analysis. A Model for the Study of Small Groups*. Repr. ed. University of Chicago Press, 1976.

——. "Social Interaction." RAND paper P-587, 1954.

——. "Some Statistical Problems in Small Group Research." *Journal of the American Statistical Association* 46, no. 255 (September 1951): 311–22.

Bales, Robert Freed, Merrill M. Flood, and A. S. Householder. "Some Interaction Models." RAND research memorandum RM-953, December 14, 1952.

Bales, Robert Freed, and Fred L. Strodtbeck. "Phases in Group Problem-Solving." *Journal of Abnormal and Social Psychology* 46 (1951): 482–95.

Baritz, Loren. *Servants of Power: A History of the Use of Social Science in American Industry*. Middletown, CT: Wesleyan University Press, 1960.

Barnett, Homer G. *Anthropology in Administration*. Evanston, IL: Row, Peterson and Company, 1956.

Bellman, Richard. *Eye of the Hurricane: An Autobiography*. Singapore: World Scientific, 1984.

Berlinski, David. *The Advent of the Algorithm: The 300-Year Journey from an Idea to the Computer.* New York: Harcourt, 2000.

Berrein, F. Kenneth. "Shelter Owners, Dissonance and the Arms Race." *Social Problems* 11 (1963): 87–91.

Birnbaum, Michael H. "Base Rates in Bayesian Inference: Signal Detection Analysis of the Cab Problem." *American Journal of Psychology* 96 (1983): 85–94.

Black, Duncan. *The Theory of Committees and Elections.* Cambridge: Cambridge University Press, 1958.

Blight, James G., and David A. Welch. *On the Brink: Americans and Soviets Reexamine the Cuban Missile Crisis.* New York: Hill and Wang, 1989.

Bluma, Lars. *Norbert Wiener und die Entstehung der Kybernetik im zweiten Weltkrieg.* Münster: Lit, 2005.

Bockman, Johanna, and Michael A. Bernstein. "Scientific Community in a Divided World: Economists, Planning, and Research Priority during the Cold War." *Comparative Studies in Society and History* 50 (2008): 581–613.

Boden, Margaret A. *Mind as Machine: A History of Cognitive Science.* 2 vols. New York: Oxford University Press, 2006.

Boring, Alan. "Computer System Reliability and Nuclear War." *Communications of the ACM* 30 (1987): 112–31.

Botterill, George, and Peter Carruthers. *The Philosophy of Psychology.* Cambridge: Cambridge University Press, 1999.

Boulding, Kenneth E. *The Image: Knowledge in Life and Society.* Ann Arbor: University of Michigan Press, 1956.

Bozeman, Barry. *Bureaucracy and Red Tape.* Upper Saddle River, NJ: Prentice Hall, 2000.

Bradley, David. *No Place to Hide.* Boston: Little Brown, 1948.

Bredekamp, Horst. *Antikensehnsucht und Maschinenglauben. Die Geschichte der Kunstkammer und die Zukunft der Kunstgeschichte.* Berlin: Klaus Wagenbach, 1993.

Brehm, Jack W., and Arthur R. Cohen. *Explorations in Cognitive Dissonance.* New York: John Wiley & Sons, 1962.

Bronfenbrenner, Urie. "The Mirror Image in Soviet-American Relations: A Social Psychologist's Report." *Journal of Social Issues* 17 (1961): 45–56.

Brown, George W. "History of RAND's Random Digits—Summary." RAND paper P-113, June 1949.

Bryant, Peter [Peter George]. *Red Alert.* Rockville, MD: Black Mask, [1958] 2008.

Butsch, Richard. *The Citizen Audience: Crowds, Publics, and Individuals.* New York: Routledge, 2008.

Byrne, Peter. *The Many Worlds of Hugh Everett III: Multiple Universes, Mutual Assured Destruction, and the Meltdown of a Nuclear Family.* Oxford: Oxford University Press, 2010.

Capshew, James. "Psychologists on Site: A Reconnaissance of the Historiography of the Laboratory." *American Psychologist* 47 (1992): 132–42.

Casscells, Ward, Arno Schoenberger, and Thomas B. Grayboys. "Interpretation by Physicians of Clinical Laboratory Results." *New England Journal of Medicine* 299 (1978): 999–1001.

Charnes, Abraham. "Optimality and Degeneracy in Linear Programming." *Econometrica* 20, no. 2 (1952): 160–70.

Charnes, Abraham, William Cooper, and Bob Mellon. "Blending Aviation Gasolines—A Study in Programming Interdependent Activities." Paper read at Project SCOOP Symposium on Linear Inequalities and Programming, June 14–16, 1951.

Christensen-Szalanski, Jay J., and Lee Roy Beach. "The Citation Bias: Fad and Fashion in the Judgment and Decision Literature." *American Psychologist* 39 (1984): 75–78.

Clemens, Walter C., Jr. "GRIT at Panmunjom: Conflict and Cooperation in a Divided Korea." *Asian Survey* 13 (1973): 531–59.

Cochran, Molly, ed. *The Cambridge Companion to Dewey*. Cambridge: Cambridge, University Press, 2010.

Cohen, L. Jonathan. "Bayesianism versus Baconianism in the Evaluation of Medical Diagnostics." *British Journal for the Philosophy of Science* 31 (1980): 45–62.

———. "Can Human Irrationality Be Experimentally Demonstrated?" *Behavioral and Brain Sciences* 4 (1981): 317–331 (comments and responses, 331–359).

———. "On the Psychology of Prediction: Whose is the Fallacy?" *Cognition* 7 (1979): 385–407.

———. *The Probable and the Provable*. Oxford: Clarendon Press, 1977.

———. "Some Historical Remarks on the Baconian Concept of Probability." *Journal of the History of Ideas* 41 (1980): 219–31.

———. "Whose is the Fallacy? A Rejoinder to Kahneman and Tversky." *Cognition* 8 (1980): 89–92.

Cohen-Cole, Jamie. "Instituting the Science of Mind: Intellectual Economies and Disciplinary Exchange at Harvard's Center for Cognitive Studies." *British Journal for the History of Science* 40 (2007): 567–97.

———. "The Creative American: Cold War Salons, Social Science, and the Cure for Modern Society." *Isis* 100, no. 2 (2009): 219–62.

Colebrooke, Henry Thomas. "Address on Presenting the Gold Medal of the Astronomical Society to Charles Babbage." *Memoirs of the Astronomical Society* 1 (1852): 509–12.

Collins, H. M. *Artificial Experts: Social Knowledge and Intelligent Machines*. Cambridge, MA: MIT Press, 1990.

Collins, Martin J. *Cold War Laboratory: RAND, the Air Force, and the American State, 1945–1950*. Washington DC: Smithsonian Institution Press, 2002.

Condorcet, M. J. A. N. "Élémens d'arithmétique et de géométrie." *Enfance* 4 (1989): 40–58.

———. *Esquisse d'un tableau historique de progrès de l'esprit humain*. Edited by O. H. Prior, édition présentée par Yvon Belaval. Paris: Librairie philosophique J. Vrin, 1970.

———. *Essai sur l'application de l'analyse à la pluralité des décisions rendues à la pluralité des voix*. Paris: Imprimerie Royale, 1785.

———. "Moyens d'apprendre à compter surement et avec facilité." *Enfance* 4 (1989): 59–90.

———. "Vie de Turgot." In *Oeuvres de Condorcet*, edited by François Arago and Arthur Condorcet-O'Connor, vol. 5, 159–60. Paris: F. Didot Frères, 1847.

Converse, Elizabeth. "The War of All against All: A Review of *The Journal of Conflict Resolution, 1957–1968*." *Journal of Conflict Resolution* 12 (1968): 471–532.

Conway, Flo, and Jim Siegelman. *Dark Hero of the Information Age: In Search of Norbert Wiener, the Father of Cybernetics*. New York: Basic Books, 2005.

Cook, William J. *The Pursuit of the Traveling Salesman: Mathematics at the Limits of Computation*. Princeton, NJ: Princeton University Press, 2012.

Cooper, William W., Abraham Charnes, and A. Henderson. *An Introduction to Linear Programming*. New York: Wiley, 1953.

Cosmides, Leda. "The Logic of Social Exchange: Has Natural Selection Shaped How Humans Reason? Studies with the Wason Selection Task." *Cognition* 31 (1989): 187–276.

Cosmides, Leda, and John Tooby. "Cognitive Adaptations for Social Exchange." In *The Adapted Mind*, edited by Jerome H. Barkow, Leda Cosmides, and John Tooby, 163–228. Oxford: Oxford University Press, 1992.

Cowles Commission for Research in Economics. *Report for Period January 1, 1948–June 30, 1949.* http://cowles.econ.yale.edu/P/reports/1948–49.htm.

———. *Rational Decision-Making and Economic Behavior. 19th Annual Report, July 1, 1950–June 30, 1951.* http://cowles.econ.yale.edu/P/reports/1950–51.htm.

Crook, Paul. *Darwinism, War, and History: The Debate over the Biology of War from the "Origin of Species" to the First World War.* Cambridge: Cambridge University Press, 1994.

Crowther-Heyck, Hunter. *Herbert A. Simon: The Bounds of Reason in Modern America.* Baltimore, MD: Johns Hopkins University Press, 2005.

Dantzig, George B. "Concepts and Origins of Linear Programming." RAND paper P-980, 1957.

———. "The Diet Problem." *Interfaces* 20 (1990): 43–47.

———. "Impact of Linear Programming on Computer Development." *OR/MS Today* 15 (1988): 12–17.

———. "Linear Programming." In *History of Mathematical Programming: A Collection of Personal Reminiscences*, edited by Jan Lenstra, Alexander Kan, and Alexander Schrijver, 19–31. Amsterdam: CWI, 1991.

———. *Linear Programming and Extensions.* Princeton, NJ: Princeton University Press, 1963.

———. "Management Science in the World of Today and Tomorrow." *Management Science* 13 (1976): C107–11.

———. "Programming of Interdependent Activities: II Mathematical Model." *Econometrica* 17, no. 3/4 (1949): 200–11.

———. "A Proof of the Equivalence of the Programming Problem and the Game Problem." In *Activity Analysis*, edited by T. C. Koopmans, 330–35. New York: John Wiley & Sons, 1951.

———. "Reminiscences about the Origins of Linear Programming." *Operations Research Letters* 1 (1982): 43–48.

Daston, Lorraine. *Classical Probability in the Enlightenment.* Princeton, NJ: Princeton University Press, 1988.

———. "Enlightenment Calculations." *Critical Inquiry* 21 (1994): 182–202.

Davis, Martin. *The Universal Computer: The Road from Leibniz to Turing.* New York: W. W. Norton, 2000.

Davis, Paul K., and John Arquilla. "Thinking about Opponent Behavior in Crisis and Conflict: A Generic Model for Analysis and Group Discussion." RAND note AD–A253 258, July 29, 1992.

Deutsch, Morton. "Conditions Affecting Cooperation." Final technical report for the Office of Naval Research, Contract NONR-285[10], February 1957.

Deutsch, Morton, and Mary Evans Collins. *Interracial Housing: A Psychological Evaluation of a Social Experiment.* Minneapolis: The University of Minnesota Press, 1951.

Dever, Gregory. "Ebeye, Marshall Islands: A Public Health Hazard." Honolulu, Hawaii: Micronesia Support Committee, 1978.

Diderot, Denis, and Jean d'Alembert, eds. *Encyclopédie, ou Dictionnaire raisonné des sciences, des arts et des métiers.* Lausanne/Berne: Les sociétés typographiques, 1780.

Dilthey, Wilhelm. *Einleitung in die Geisteswissenschaften* [1883]. In *Wilhelm Dilthey, Gesam-*

melte Schriften, edited by Bernhard Groethuysen, vol. 1. Stuttgart/Göttingen: B.G. Teubner/Vandenhoeck & Ruprecht, 1990.

Dimand, Robert W., and Mary Ann Dimand. "The Early History of the Strategic Theory of Games from Waldgrave to Borel." In *Toward a History of Game Theory*, edited by E. Roy Weintraub, 15–28. Durham, NC: Duke University Press, 1992.

Dodge, Robert. *The Strategist: The Life and Times of Thomas Schelling*. Hollis, NH: Hollis Publishing, 2006.

Dreyfus, Hubert. *What Computers Can't Do: A Critique of Artificial Reason*. New York: Harper and Row, 1972.

Dubin, Robert. "Review of *Patterns of Industrial Bureaucracy*, by Gouldner." *American Sociological Review* 20 (1955): 120–22.

Dunaway, Edward. U.S. Air Force Oral History. Interview by Daniel R. Mortensen, April 17, 1980, Transcript, Office of Air Force History IRIS No. 01129703.

———. U.S. Air Force Oral History. Interview by James R. Luntzel, June 22, 1973, Transcript, Office of Air Force History IRIS No. 01129703.

"Economics and Operations Research: A Symposium." *Review of Economics and Statistics* 40 (1958): 195–229.

Edwards, Paul. N. *The Closed World: Computers and the Politics of Discourse in Cold War America*. Cambridge, MA: MIT Press, 1996.

Edwards, Ward. "Behavioral Decision Theory." *Annual Review of Psychology* 12 (1961): 473–98.

Ellsberg, Daniel. "A Final Comment." *Review of Economics and Statistics* 40, no. 3 (1958): 227–29.

———. "Risk, Ambiguity, and the Savage Axioms." *Quarterly Journal of Economics* 75, no. 4 (1961): 643–69.

———. "The Theory and Practice of Blackmail." RAND Corporation Report, Santa Monica, CA, July 1968, 1–38, http://www.rand.org/content/dam/rand/pubs/papers/2005/P3883.pdf.

Elster, Jon. *Ulysses and the Sirens: Studies in Rationality and Irrationality*. Cambridge: Cambridge University Press, [1979] 1990.

Engerman, David. *Know Your Enemy: The Rise and Fall of America's Soviet Experts*. New York: Oxford University Press, 2009.

Engerman, David. "Social Science in the Cold War." *Isis* 101 (2010): 393–400.

Enke, Stephen. "Some Economic Aspects of Fissionable Material." *Quarterly Journal of Economics* 68 (1954): 217–32.

Erickson, Paul. "Optimism and Optimization," In Erickson, *The World the Game Theorists Made*. Unpublished manuscript.

Etzioni, Amitai. "The Kennedy Experiment." *Western Political Quarterly* 20 (1967): 361–80.

Evans, Jonathan St. B. T. "Reasoning with Bounded Rationality." *Theory & Psychology* 2 (1992): 237–42.

———. "Theories of Human Reasoning: The Fragmented State of the Art." *Theory & Psychology* 1 (1991): 83–105.

Evans, Jonathan St. B. T., and David E. Over. *Rationality and Reasoning*. Hove, UK: Psychology Press, 1996.

Festinger, Leon. *A Theory of Cognitive Dissonance*. Stanford, CA: Stanford University Press, 1957.

Festinger, Leon, Henry W. Riecken, and Stanley Schachter. *When Prophecy Fails: A Social and Psychological Study of a Modern Group that Predicted the Destruction of the World*. New York: Harper Torchbooks, 1956.

Fiedler, Klaus. "The Dependence of the Conjunction Fallacy on Subtle Linguistic Factors." *Psychological Research* 50 (1988): 123–29.

Finetti, Bruno de. "La prévision: Ses lois logiques, ses sources subjectives." *Annales de l'Institute Henri Poincaré* 7 (1937): 1–68.

Firth, Stewart. *Nuclear Playground: Fight for an Independent and Nuclear Free Pacific*. Honolulu: University of Hawai'i Press, 1987.

Fischhoff, Baruch, Paul Slovic, Sarah Lichtenstein, Stephen Read, and Barbara Combs. "How Safe Is Safe Enough? A Psychometric Study of Attitudes towards Technological Risks and Benefits." *Policy Sciences* 9 (1978): 127–52.

Fisher, Paul. "The Berlin Airlift." *The Beehive, United Aircraft Corporation* 23 (1948): 1–31.

Fleck, Ludwik. *Genesis and Development of a Scientific Fact*. Translated by F. Bradley and T. Trenn. Chicago: University of Chicago Press, [1935] 1979.

Flood, Merrill M. "The Objectives of TIMS." *Management Science* 2 (1956): 178–84.

———. "Some Experimental Games." RAND research memorandum RM-789-1, June 20, 1952.

Flowers, Matie L. "A Laboratory Test of Some Implications of Janis's Groupthink Hypothesis." *Journal of Personality and Social Psychology* 35 (1977): 888–96.

Foster, George M. *Long-Term Field Research in Social Anthropology*. London: Academic Press, 1979.

Freedman, Lawrence. *The Evolution of Nuclear Strategy*. New York: Palgrave Macmillan, [1981] 2003.

Frydenberg, Erica. *Morton Deutsch: A Life and Legacy of Mediation and Conflict Resolution*. Brisbaine: Australian Academic Press, 2005.

Gaddis, John Lewis. "International Relations Theory and the End of the Cold War." *International Security* 17 (1992–1993): 5–58.

Gale, David, Harold W. Kuhn, and Albert W. Tucker. "Linear Programming and the Theory of Games." In *Activity Analysis*, edited by T. C. Koopmans, 317–29. New York: John Wiley & Sons, 1951.

Galison, Peter. *Image and Logic: A Material Culture of Microphysics*. Chicago: University of Chicago Press, 1997.

———. "The Ontology of the Enemy: Norbert Wiener and the Cybernetic Vision." *Critical Inquiry* 21(1994): 228–66.

Garfinkel, Harold. "The Rational Properties of Scientific and Common Sense Activities." *Behavioral Sciences* 5 (1962): 72–83.

Gass, Saul I. "Model World: In the Beginning There Was Linear Programming." *Interfaces* 20 (1990): 128–32.

Gass, Saul I., and Arjang Assad. *Annotated Timeline of Operations Research*. New York: Springer, 2005.

Geertz, Clifford. *The Interpretation of Cultures: Selected Essays*. New York: Basic Books, 1973.

Geisler, Murray A. *A Personal History of Logistics*. Bethesda, MD: Logistics Management Institute, 1986.

Gerovitch, Slava. *From Newspeak to Cyberspeak: A History of Soviet Cybernetics*. Cambridge, MA: MIT Press, 2002.

Ghamari-Tabrizi, Sharon. *The Worlds of Herman Kahn: The Intuitive Science of Thermonuclear War*. Cambridge, MA: Harvard University Press, 2005.

Gigerenzer, Gerd. *Adaptive Thinking*. New York: Oxford University Press, 2000.

Gigerenzer, Gerd. "Striking a Blow for Sanity in Theories of Rationality." In *Models of a Man: Essays in Memory of Herbert A. Simon*, edited by Mie Augier and James G. March, 389–409. Cambridge, MA: MIT Press, 2004.

Gigerenzer, Gerd, Zeno Swijtink, Theodore Porter, Lorraine Daston, John Beatty, and Lorenz Krüger. *The Empire of Chance*. Cambridge: Cambridge University Press, 1989.

Gilbreth Carey, Ernestine, and Frank B. Gilbreth. *Cheaper by the Dozen*. New York: Crowell, 1948.

Gillespie, Richard. *Manufacturing Knowledge: A History of the Hawthorne Experiments*. Cambridge: Cambridge University Press, 1993.

Gilovich, Thomas. *How We Know What Isn't So: The Fallibility of Human Reason in Everyday Life*. New York: Free Press, 1991.

Gilovich, Thomas, Dale W. Griffin, and Daniel Kahneman, eds. *Heuristics and Biases: The Psychology of Intuitive Judgment*. Cambridge: Cambridge University Press, 2002.

Giocoli, Nicola. *Modeling Rational Agents from Interwar Economics to Early Modern Game Theory*. Cheltenham, UK: Edward Elgar, 2003.

Goodman, Nelson, *Fact, Fiction, and Forecast*. London: Athlone, 1954.

Gordin, Michael D. *Red Cloud at Dawn: Truman, Stalin, and the End of the Atomic Monopoly*. New York: Farrar, Straus and Giroux, 2009.

Gould, Julius, and W. L. Kolb, eds. *A Dictionary of the Social Sciences*. Glencoe, IL: Free Press, 1964, 573–74.

Gouldner, Alvin W. *Patterns of Industrial Bureaucracy*. Glencoe, IL: Free Press, 1954.

Graebner, William. *The Engineering of Consent: Democracy and Authority in Twentieth-Century America*. Madison: University of Wisconsin Press, 1987.

Granger, Gilles-Gaston. *La Mathématique sociale du marquis de Condorcet*. Paris: Presses universitaires de France, 1956.

Grattan-Guinness, Ivor. *The Search for Mathematical Roots, 1870–1940: Logic, Set Theories, and the Foundations of Mathematics from Cantor through Russell to Gödel*. Princeton, NJ: Princeton University Press, 2000.

Grier, David Alan. *When Computers Were Human*. Princeton, NJ: Princeton University Press, 2006.

Griggs, Richard, and James R. Cox. "The Elusive Thematic Materials Effects in Wason's Selection Task." *British Journal of Psychology* 73 (1982): 407–20.

Grimm, Jacob, and Wilhelm Grimm. *Deutsches Wörterbuch*. Leipzig: S. Hirzel, 1893; facsimile repr., 1991.

Grunewald, Robert N. "Dewey's 'Situation' and the Ames Demonstrations." *Educational Theory* 15, no. 4 (1965): 293–304.

Gunther, John. "What Do You Know About You?" *Look*, May 9, 1950.

Haag, Johannes. "Regel." In *Kant-Lexikon*, edited by Georg Mohr, Jürgen Stolzenberg, and Marcus Willaschek. Berlin: De Gruyter, forthcoming.

Haas, Mark L. "Prospect Theory and the Cuban Missile Crisis." *International Studies Quarterly* 45 (2001): 241–70.

Hacking, Ian. "Inaugural Lecture for the Chair of Philosophy at the Collège de France." *Economy and Society* 31, no. 1 (2001): 1–14.

Hagen, Everett E. "Analytical Models in the Study of Social Systems." *American Journal of Sociology* 67 (1961): 144–51.

Haldeman, Harry R., and Joseph DiMona. *The Ends of Power*. New York: Times Books, 1978.

Hall, Elizabeth. "A Conversation with Charles Osgood." *Psychology Today*, November 1973, 54–56, 58–60, 64–72.

Hamilton, William D. "Extraordinary Sex Ratios." *Science* 156.3774, new series (April 28, 1967): 477–88.

———. "The Evolution of Reciprocal Altruism." *Quarterly Review of Biology* 46 (1971): 35–57.

———. "The Genetical Evolution of Social Behavior, I." *Journal of Theoretical Biology* 7 (1964): 1–26.

———. "The Genetical Evolution of Social Behavior, II." *Journal of Theoretical Biology* 7 (1964): 27–52.

———. "Selection of Selfish and Altruistic Behavior in Some Extreme Models." In *Man and Beast: Comparative Social Behavior,* edited by J. F. Eisenberg and Wilton S. Dillon, 59–91. Washington DC: Smithsonian Institution Press, 1971.

Hammond, Deborah. "Toward a Systems View of Democracy: The Society for General Systems Research, 1954–1978." PhD diss., University of California, Berkeley, 1997.

Harlow, Harry. "The Nature of Love." *American Psychologist* 13 (1958): 673–85.

Hart, Paul 't. *Groupthink in Government: A Study of Small Groups and Policy Failure.* Baltimore, MD: Johns Hopkins University Press, 1990.

———. "Irving L. Janis' Victims of Groupthink." *Political Psychology* 12 (1991): 247–78.

Hart, Paul 't, Eric K. Stern, and Bengt Sundelius, eds. *Beyond Groupthink: Political Group Dynamics and Foreign Policy-Making.* Ann Arbor: University of Michigan Press, 1997.

Harty, Martha, and John Modell. "The First Conflict Resolution Movement, 1956–1971: An Attempt to Institutionalize Applied Interdisciplinary Social Science." *Journal of Conflict Resolution* 35 (1991): 720–58.

Heims, Steve J. *The Cybernetics Group.* Cambridge, MA: MIT Press, 1991.

Helmer, Olaf. "Recent Developments in the Mathematical Theory of Games." (RAOP-16, 30 April 1948): 16–18.

———. "Strategic Gaming." RAND paper P-1902, 1960.

Henderson, William, and B. W. Aginsky. "A Social Science Field Laboratory." *American Sociological Review* 6 (1941): 41–44.

Henriksen, Margot A. *Dr. Strangelove's America: Society and Culture in the Atomic Age.* Berkeley: University of California Press, 1997.

Herman, Elle. *The Romance of American Psychology: Political Culture in the Age of Experts.* Berkeley: University of California Press, 1995.

Heukelom, Floris. "Kahneman and Tversky and the Making of Behavioral Economics." PhD diss., University of Amsterdam, 2009.

Heyck, Hunter. "Producing Reason." In *Cold War Social Science: Production, Liberal Democracy, and Human Nature,* edited by Mark Solovey and Hamilton Cravens, 99–116. New York: Palgrave, 2011.

Heyck, Hunter, and David Kaiser. "Introduction: New Perspectives on Science and the Cold War." *Isis* 101 (2010): 362–66.

Hilbert, David, and Wilhelm Ackermann. *Grundzüge der theoretischen Logik.* Berlin: Springer, 1928.

Hoffmann, Ludwig. *Mathematisches Wörterbuch,* 7 vols. Berlin: Wiegandt & Hempel, 1858–1867.

Holmberg, Allan. "Experimental Intervention in the Field." In *Peasants, Power, and Applied Social Change: Vicos as a Model,* edited by Henry F. Dobyns, Paul L. Doughty, and Harold Lasswell, 33–64. London: Sage, 1971.

Hontheim, Joseph. *Der logische Algorithmus in seinem Wesen, in seiner Anwendung und in seiner philosophischen Bedeutung.* Berlin: Felix L. Dames, 1895.

Hopmann, P. Terrence, and Timothy King. "Interactions and Perceptions in the Test Ban Negotiations." *International Studies Quarterly* 20 (1976): 105–42.

Hounshell, David A. *From the American System to Mass Production, 1800–1932*. Baltimore, MD: Johns Hopkins University Press, 1984.

Howes, Davis, and Charles E. Osgood. "On the Combination of Associative Probabilities in Linguistic Contexts." *American Journal of Psychology* 67 (1954): 241–58.

Husbanks, Philip, Owen Holland, and Michael Wheeler, eds. *The Mechanical Mind in History*. Cambridge, MA: MIT Press, 2008.

Huxley, Julian. "Introduction: A Discussion of Ritualization of Behavior in Animals and Man." *Philosophical Transactions of the Royal Society of London Series B (Biological Sciences)* 251.772 (1966): 249–71.

Immerman, Richard H. "Psychology." *Journal of American History* 77 (1990): 169–80.

Inhelder, Bärbel, and Jean Piaget. *De la logique de l'enfant à la logique de l'adolescence*. Paris: PUF, 1955.

———. *The Growth of Logical Thinking from Childhood to Adolescence*. New York: Basic Books, 1958.

Israel, Giorgio. *The World as Mathematical Game: John von Neumann and Twentieth-Century Science*. Basel: Birkhäuser, 2009.

Jacobs, Walter. "Air Force Progress in Logistics Planning." *Management Science* 3 (1957): 213–24.

Janis, Irving L. *Groupthink: Psychological Studies of Policy Decisions and Fiascoes*. Boston: Houghton Mifflin Company, 1983.

———. *Victims of Groupthink: A Psychological Study of Foreign-Policy Decisions and Fiascoes*. Boston: Houghton Mifflin Company, 1972.

Johnson, Lyle R. "Coming to Grips with Univac." *IEEE Annals of the History of Computing Archive* 28 (2006): 32–42.

Johnson-Laird, Philip. "Peter Wason: Obituary." *Guardian*, April 25, 2003.

Joseph, Alice, and Veronica Murray. *Chamorros and Carolinians of Saipan: Personality Studies*. Cambridge, MA: Harvard University Press, 1951.

Juda, Lawrence. "Negotiating a Treaty on Environmental Modification Warfare: The Convention on Environmental Warfare and Its Impact upon Arms Control Negotiations." *International Organization* 32 (1978): 975–91.

Jung, Matthias. "John Dewey and Action." In *The Cambridge Companion to Dewey*, edited by Molly Cochran, 145–65. Cambridge: Cambridge, University Press, 2010.

Jungermann, Helmut. "The Two Camps on Rationality." In *Decision Making under Uncertainty*, edited by Roland W. Scholz, 63–86. Amsterdam: North-Holland, 1983.

Kahn, Herman. *On Escalation: Metaphors and Scenarios*. New York: Frederick A. Praeger, 1965.

———. *On Thermonuclear War*. Princeton, NJ: Princeton University Press, 1960.

Kahneman, Daniel. "Autobiography." Nobelprize.org. 2012. http://nobelprize.org/nobel _prizes/economics/laureates/2002/kahneman-autobio.html.

———. "Maps of Bounded Rationality: Psychology for Behavioral Economics." *American Economic Review* 93 (2003): 1449–75.

Kahneman, Daniel, Paul Slovic, and Amos Tversky, eds. *Judgment under Uncertainty: Heuristics and Biases*. New York, 1982.

Kahneman, Daniel, and Amos Tversky. "On the Interpretation of Intuitive Probability: A Reply to Jonathan Cohen." *Cognition* 7 (1980): 409–11.

———. "On the Psychology of Prediction." *Psychological Review* 80 (1973): 237–51.

———. "Prospect Theory: An Analysis of Decision under Risk." *Econometria* 47 (1979): 263–92.

———. "Subjective Probability: A Judgment of Representativeness." *Cognitive Psychology* 3 (1972): 430–54.

Kalisch, G. et al. "Some Experimental N-Person Games." RAND research memorandum RM-948, 1952.

Kant, Immanuel. *Critique of Judgment*. Translated by Werner S. Pluhar. Indianapolis, IN: Hackett, 1987.

———. *Critique of Pure Reason*. Translated and edited by Paul Guyer and Allen W. Wood. Cambridge: Cambridge University Press, 1997.

Kantorovich, Leonid. "Mathematical Methods of Organizing and Planning Production." *Management Science* 6, no. 4 (1960): 366–422.

Kanwisher, Nancy. "Cognitive Heuristics and American Security Policy." *Journal of Conflict Resolution* 33 (1989): 652–75.

Kaplan, Bert, ed. *Primary Records in Culture and Personality*. Vol. 1. Madison, WI: Microcard Foundation, 1957.

Kaplan, Fred M. *The Wizards of Armageddon*. New York: Simon and Schuster, 1983.

Kay, Lily E. *Who Wrote the Book of Life? A History of the Genetic Code*. Stanford, CA: Stanford University Press, 2000.

Keesing, Felix M. "Administration in Pacific Islands." *Far Eastern Survey* 16 (1947): 61–65.

Keller, Evelyn Fox. *The Century of the Gene*. Cambridge, MA: Harvard University Press, 2000.

Kennan, George F. *The Nuclear Delusion: Soviet-American Relations in the Atomic Age*. New York: Pantheon, 1983.

Kennedy, Robert F. *Thirteen Days: A Memoir of the Cuban Missile Crisis*. New York: W. W. Norton, [1968] 1971.

Kevles, Daniel J. *In the Name of Eugenics: Genetics and the Uses of Human Heredity*. New York: Knopf, 1985.

Kissinger, Henry A. *Nuclear Weapons and Foreign Policy*. New York: Harper & Brothers, 1957.

Kiste, Robert, and Mac Marshall. "American Anthropology in Micronesia 1947–1991." *Pacific Science* 54 (2000): 265–74.

Klaes, Matthias, and Esther-Mirjam Sent. "The Conceptual History of the Emergence of Bounded Rationality." *History of Political Economy* 37, no. 1 (2005): 27–59.

Klein, Judy L. "Reflections from the Age of Measurement." In *The Age of Economic Measurement*, edited by Judy L. Klein and Mary S. Morgan, 128–33. Durham, NC: Duke University Press, 2001.

———. *Protocols of War and the Mathematical Invasion of Policy Space, 1940–1960*. Unpublished manuscript, 2012. MS Word document.

Kluckhohn, Clyde. "A Comparative Study of Values in Five Cultures." In "Navaho Veterans," by Evon Vogt. *Papers of the Peabody Museum of Harvard University* 41, no. 1 (1951), vii–xii.

———. "The Personal Document in Anthropological Science." In *The Use of Personal Documents in History, Anthropology and Sociology*, edited by Louis R. Gottschalk, Clyde Kluckhohn, and Robert Angell, 78–193. New York: Social Science Research Council, 1945.

Kneale, William, and Martha Kneale. *The Development of Logic*. Oxford: Clarendon Press, 1962.

Kohli, Martin C. "Leontief and the U.S. Bureau of Labor Statistics, 1941–1954: Developing a framework for measurement." In *The Age of Economic Measurement*, edited by

Judy L. Klein and Mary S. Morgan, 190–212. Durham, NC: Duke University Press 2001.

Koopmans, Tjalling C., ed. *Activity Analysis of Production and Allocation: Proceedings of a Conference*. Cowles Commission for Research in Economics Monograph no. 13. New York: John Wiley & Sons, 1951.

——. Autobiography. Nobelprize.org. http://nobelprize.org/nobel_prizes/economics/laureates/1975/koopmans.html.

——. "Concepts of Optimality and Their Uses." Nobel memorial lecture, December 11, 1975. http:// nobelprize.org/nobel_prizes/economics/laureates/1975/koopmans-lecture.pdf.

——. Introduction to *Activity Analysis of Production and Allocation: Proceedings of a Conference*. Cowles Commission for Research in Economics Monograph no. 13, edited by T. C. Koopmans, 1–12. New York: John Wiley & Sons, 1951.

——. "A Note about Kantorovich's Paper, 'Mathematical Methods of Organizing and Planning Production.'" *Management Science* 6 (1960): 363–65.

Krause, George A., and Kenneth J. Meier, eds. *Politics, Policy, and Organizations: Frontiers in the Scientific Study of Bureaucracy*. Ann Arbor: University of Michigan Press, 2006.

Kreps, David M., Paul Milgrom, John Roberts, and Robert Wilson. "Rational Cooperation in the Finitely Repeated Prisoners' Dilemma." *Journal of Economic Theory* 27, no. 2 (1982): 245–52.

Labedz, Leopold. *Poland under Jaruzelski: A Comprehensive Sourcebook on Poland during and after Martial Law*. New York: Scribner, 1984.

Langer, Ellen. *The Psychology of Control*. Beverly Hills, CA: Sage Publications, 1983.

Lasch, Christopher. "The Social Theory of the Therapeutic: Parsons and the Parsonians." In *Haven in a Heartless World: The Family Besieged*, 111–33. New York: Norton, 1977.

Lawrence, Douglas H., and Leon Festinger. *Deterrents and Reinforcement: The Psychology of Insufficient Reward*. Stanford, CA: Stanford University Press, 1962.

Lazarsfeld, Paul. "The Use of Panels in Social Research." *Proceedings of the American Philosophical Society* 92 (1948): 405–10.

Lazarsfeld, Paul, and Morris Rosenberg. *The Language of Social Research*. New York: Free Press, 1955.

Lebow, Richard Ned. "The Cuban Missile Crisis: Reading the Lessons Correctly." *Political Science Quarterly* 98 (1983): 431–58.

Leffler, Melvyn P., and Odd Arne Westad. *The Cambridge History of the Cold War*. 3 vols. Cambridge: Cambridge University Press, 2010.

Leibniz, Gottfried Wilhelm. "Preface to the General Science." In *Leibniz Selections*, edited by Philip Wiener, 12–17. New York: Charles Scribner's Sons, 1951.

——. "Towards a Universal Characteristic." In *Leibniz Selections*, edited by Philip Wiener, 17–25. New York: Charles Scribner's Sons, 1951.

Leighton, Alexander. *The Effects of Atomic Bombs on Hiroshima and Nagasaki: The U.S. Strategic Bombing Survey*. Washington DC: Government Printing Office, June 30, 1946.

——. *The Governing of Men*. Princeton, NJ: Princeton University Press, 1945.

——. *Human Relations in a Changing World: Observations on the Use of the Social Sciences*. New York: Dutton, 1949.

Lenstra, J. K., A. H. G. Rinnooy Kann, and A. Schrijver, eds. *History of Mathematical Programming: A Collection of Personal Reminiscences*. Amsterdam: North-Holland, 1991.

Leonard, Robert. "'Between Worlds,' or an Imagined Reminiscence by Oskar Morgenstern about Equilibrium and Mathematics in the 1920s." *Journal of the History of Economic Thought* 26 (2004): 285–310.

———. *Von Neumann, Morgenstern, and the Creation of Game Theory: From Chess to Social Science, 1900–1960*. Cambridge: Cambridge University Press, 2010.

Leslie, Stuart W. *The Cold War and American Science: The Military-Industrial-Academic Complex at MIT and Stanford*. New York: Columbia University Press, 1993.

Levy, Ariel S. and Glen Whyte. "A Cross-Cultural Explanation of the Reference Dependence of Crucial Group Decision under Risk: Japan's 1941 Decision for War." *Journal of Conflict Resolution* 41 (1997): 792–813.

Lewin, Kurt, and Ronald Lippitt. "An Experimental Approach to the Study of Autocracy and Democracy: A Preliminary Note." *Sociometry* 1.3/4 (January–April 1938): 292–300.

Lewin, Kurt, Ronald Lippitt, and Ralph K. White. "Patterns of Aggressive Behavior in Experimentally Created 'Social Climates.'" *Journal of Social Psychology* 10 (1939): 271–99.

Lindskold, Svenn. "Trust Development, the GRIT Proposal, and the Effects of Conciliatory Acts on Conflict and Cooperation." *Psychological Bulletin* 85 (1978): 772–93.

Lopes, Lola L. "Performing Competently." *Behavioral and Brain Sciences* 4 (1981): 343–44.

———. "The Rhetoric of Irrationality." *Theory & Psychology* 1 (1991): 65–82.

Lucas, William F. "The Proof that a Game May Not Have a Solution." RAND research memo RM-5543-PR, January 1968.

Luce, R. Duncan, and Howard Raiffa. *Games and Decisions: Introduction and Critical Survey*. Mineola, NY: Dover, [1957] 1985.

Maas, Harro. *William Stanley Jevons and the Making of Modern Economics*. Cambridge: Cambridge University Press, 2005.

MacMartin, Clare, and Andrew Winston. "The Rhetoric of Experimental Social Psychology, 1930–1960: From Caution to Enthusiasm." *Journal of the History of the Behavioral Sciences* 36, no. 4 (2000): 349–64.

Margolis, Julius. "Discussion." In *Strategic Interaction and Conflict: Original Papers and Discussion*, edited by Kathleen Archibald, 137. Berkeley: Institute of International Studies, 1966.

Marguin, Jean. *Histoire des instruments à calculer. Trois siècles de mécanique pensante 1642–1942*. Paris: Hermann, 1994.

Markov, A. A. *Theory of Algorithms*. Translated by Jacques J. Schorr-Kon and PST Staff. Moscow: Academy of Sciences of the USSR, 1954.

Marrow, Alfred J. *The Practical Theorist: The Life and World of Kurt Lewin*. New York: Basic Books, 1969.

Marschak, Jacob. "Rational Behavior, Uncertain Prospects, and Measurable Utilities." *Econometrica* 18 (1950): 111–41.

Marten Zisk, Kimberly. "Soviet Academic Theories on International Conflict and Negotiations." *Journal of Conflict Resolution* 34 (1990): 678–93.

Mayer, Otto. *Authority, Liberty, and Automatic Machinery in Early Modern Europe*. Baltimore, MD: Johns Hopkins University Press, 1986.

Maynard Smith, John. *Evolution and the Theory of Games*. Cambridge: Cambridge University Press, 1982.

McCarthy, Anna. "'Stanley Milgram, Allen Funt, and Me': Postwar Social Science and the 'First Wave' of Reality TV." In *Reality TV: Remaking Television*, edited by Susan Murray and Laurie Ouellette, 23–44. New York: New York University Press, 2004.

McCorduck, Pamela. *Machines Who Think: A Personal Inquiry into the History and Prospects of Artificial Intelligence*. 2nd ed. Natick, MA: A. K. Peters, [1979] 2004.

McDermott, Rose. "Arms Control and the First Reagan Administration: Belief-Systems and Policy Choices." *Journal of Cold War Studies* 4 (2002): 29–59.

———. "The Psychological Ideas of Amos Tversky and Their Relevance for Political Science." *Journal of Theoretical Politics* 13 (2001): 5–33.

———. *Risk Taking in International Politics: Prospect Theory in Postwar American Foreign Policy.* Ann Arbor: University of Michigan Press, 1998.

McDonald, John. *Strategy in Poker, Business, and War.* New York: W. W. Norton & Company, 1950.

Meehl, Paul E. "Causes and Effects of My Disturbing Little Book." *Journal of Personality Assessment* 50 (1986): 370–75.

———. *Clinical versus Statistical Prediction. A Theoretical Analysis and a Review of the Literature.* Minneapolis: University of Minnesota Press, 1954.

———. "When Shall We Use Our Heads Instead of the Formula?" *Journal of Counseling Psychology* 4 (1957): 268–73.

Mehmke, R. "Numerisches Rechnen." In *Enzyklopädie der Mathematischen Wissenschaften*, edited by Wilhelm Franz Meyer. Vol. 1, part 2, 959–78. Leipzig: B.G. Teubner, 1898–1934.

Mele, Alfred R., and Piers Rawling, eds. *The Oxford Handbook of Rationality.* Oxford: Oxford University Press, 2004.

Menabrea, Luigi F. "Sketch of the Analytical Engine Invented by Charles Babbage." Translated by Ada Augusta, Countess of Lovelace, first published in the *Bibliothèque de Genève* in 1842 and reprinted in *Charles Babbage and His Calculating Engines*, edited by Philip Morrisson and Emily Morrison, 225–95. New York: Dover, 1961.

Mercer, Jonathan. "Rationality and Psychology in International Politics." *International Organization* 59 (2005): 77–106.

Merton, Robert K., Marjorie Fisk Lowenthal, and Alberta Curtis. *Mass Persuasion: The Social Psychology of a War Bond Drive.* New York: Harper and Brothers, 1946.

Metropolis, Nicholas. "The Beginning of the Monte Carlo Method." *Los Alamos Science*, special issue (1987): 125–29.

Miller, James D. "Toward a General Theory for the Behavioral Sciences." In *The State of the Social Sciences*, edited by Leonard D. White, 29–45. Chicago: University of Chicago Press, 1956.

Mirowski, Philip. *Machine Dreams: Economics Becomes a Cyborg Science.* Cambridge: Cambridge University Press, 2002.

———. "When Games Grow Deadly Serious: The Military Influence upon the Evolution of Game Theory." In *Economics and National Security: A History of their Interaction.* Annual Supplement to History of Political Economy 23, edited by Craufurd D.W. Goodwin, 227–55. Durham, NC: Duke University Press, 1991.

Mises, Ludwig von. "Economic Calculation in the Socialist Commonwealth." In *Collectivist Economic Planning; Critical Studies on the Possibilities of Socialism*, edited by Friedrich A. Hayek, 87–130. London: Routledge & Kegan Paul, [1920] 1935.

Mitman, Gregg. *The State of Nature: Ecology, Community, and American Social Thought, 1900–1950.* Chicago: University of Chicago Press, 1992.

Moorhead, Gregory. "Groupthink: Hypothesis in Need of Testing." *Group & Organization Studies* 7 (1982): 429–44.

Moorhead, Gregory, and Montanari, John R. "An Empirical Investigation of the Groupthink Phenomenon." *Human Relations* 39 (1986): 399–410.

Morgan, Mary S. "The Curious Case of the Prisoner's Dilemma: Model Situation? Exemplary Narrative?" In *Science without Laws: Model Systems, Cases, Exemplary Narratives,* edited by Angela N. H. Creager, Elizabeth Lunbeck, and M. Norton Wise, 157–85. Durham, NC: Duke University Press, 2007.

———. "Economic Man as Model: Ideal Types, Idealization and Caricatures." *Journal of the History of Economic Thought* 28 (2006): 1–27.

Morgenstern, Oskar. *The Question of National Defense*. New York: Random House, 1959.

———. *Wirtschaftsprognose. Eine Untersuchung ihrer Voraussetzungen und Möglichkeiten*. Vienna: Julius Springer, 1928.

Morris, Desmond. *The Naked Ape; A Zoologist's Study of the Human Animal*. New York: McGraw-Hill, 1967.

Morton Deutsch. "Trust and Suspicion." *Journal of Conflict Resolution* 2.4 (December 1958): 265–79.

Murdock, George P. *Outline of World Cultures*. New Haven, CT: Human Relations Area Files, 1954 [subsequent editions, 1958, 1963, 1972, 1975, 1983].

Murdock, George P., Clellan S. Ford, Alfred E. Hudson, Raymond Kennedy, Leo W. Simmons, and John W. M. Whiting. *Outline of Cultural Materials*. New Haven, CT: Cross-Cultural Survey, 1938 [subsequent editions, New Haven, CT: Human Relations Area Files, 1945, 1950, 1951, 1960, 1982, 2000].

Nagel, Thomas. *The Possibility of Altruism*. Princeton, NJ: Princeton University Press, 1970.

Nash, John C. "The (Dantzig) Simplex Method for Linear Programming." *Computing in Science and Engineering* 2 (2000): 29–31.

Nash, John F. "Non-Cooperative Games." PhD diss., Princeton University, May 1950.

Nisbett, Richard E., and Eugene Borgida. "Attribution and the Psychology of Prediction." *Journal of Personal and Social Psychology* 32 (1975): 932–43.

Nisbett, Richard E., and Lee Ross. *Human Inference: Strategies and Shortcomings of Social Judgement*. Englewood Cliffs, NJ: Prentice Hall, 1980.

Nozick, Robert. *The Nature of Rationality*. Princeton, NJ: Princeton University Press, 1993.

Oaksford, Mike, and Nick Chater. "Human Rationality and the Psychology of Reasoning: Where Do We Go from Here?" *British Journal of Psychology* 92 (2001): 193–216.

Offray de la Mettrie, Julien. *Man a Machine*. Edited by Gertrude Carman Bussey. La Salle: Open Court, 1912.

Orchard-Hays, William. "Evolution of Linear Programming Computing Techniques." *Management Science* 4 (1958): 183–90.

———. "History of the Development of LP Solvers." *Interfaces* 20 (1990): 61–73.

Orden, Alex. "LP from the '40s to the '90s." *Interfaces* 23 (1993): 2–12.

Osgood, Charles E. *An Alternative to War or Surrender*. Urbana: University of Illinois Press, [1962] 1970.

———. "Disarmament Demands GRIT." In *Toward Nuclear Disarmament and Global Security: A Search for Alternatives*, edited by Burns H. Weston, 337–44. Boulder, CO: Westview Press, 1984.

———. "Graduated Unilateral Initiatives for Peace." In *Preventing World War III: Some Proposals*, edited by Quincy Wright, William M. Evan, and Morton Deutsch, 161–77. New York: Simon and Schuster, 1962.

———. "GRIT: A Strategy for Survival in Mankind's Nuclear Age?" In *New Directions in Disarmament*, edited by William Epstein and Bernard T. Feld, 164–72. New York: Praeger, 1981.

———. *Perspective in Foreign Policy*. Palo Alto, CA: Pacific Books, 1966.

———. "The Psychologist in International Affairs." *American Psychologist* 19 (1964): 111–18.

———. "Putting the Arms Race in Reverse." *Christian Century* 79 (1962): 566–68.

———. "Questioning Some Unquestioned Assumptions about National Security." *Social Problems* 11 (1963): 6–12.

———. "Reciprocal Initiatives." In *The Liberal Papers*, edited by James Roosevelt, 155–228. Garden City, NY: Anchor Books, 1962.

———. "Reversing the Arms Race." *Progressive*, May 1962, 27–31.

———. "Statement on Psychological Aspects of International Relations." In *Psychological Dimensions of Social Interaction: Readings and Perspectives*, edited by D. E. Linder, 277–85. Reading, MA: Addison-Wesley, 1973.

———. "Suggestions for Winning the Real War with Communism." *Journal of Conflict Resolution* 3 (1959): 295–325.

Osgood, Charles E., George J. Suci, and Percy H. Tannenbaum. *The Measurement of Meaning*. Urbana: University of Illinois Press, 1957.

Osgood, Charles E., and Percy H. Tannenbaum. "The Principle of Congruity in the Prediction of Attitude Change." *Psychological Review* 62 (1955): 42–55.

Osgood, Charles E., and Oliver C. S. Tzeng. *Language, Meaning, and Culture: The Selected Papers of C. E. Osgood*. New York: Praeger, 1990.

OSS Assessment Staff. *The Assessment of Men: Selection of Personnel for the Office of Strategic Services*. New York: Rhinehart and Co., 1948.

Parfit, Derek. *Reasons and Persons*. Oxford: Oxford University Press, 1980

Peterson, Cameron R., and Lee R. Beach. "Man as an Intuitive Statistician." *Psychological Bulletin* 68 (1967): 29–46.

Piattelli-Palmarini, Massimo. *Inevitable Illusions: How Mistakes of Reason Rule Our Minds*. New York: Wiley 1994.

Pickering, Andrew. *The Cybernetic Brain: Sketches of Another Future*. Chicago: University of Chicago Press, 2010.

Pilisuk, Marc, and Paul Skolnick. "Inducing Trust: A Test of the Osgood Proposal." *Journal of Personality and Social Psychology* 8 (1968): 121–33.

Platt, Jennifer. *A History of Sociological Research Methods in America*. Cambridge: Cambridge University Press, 1999.

Popper, Karl R. *The Logic of Scientific Discovery*. New York: Basic Books, 1959.

Porter, Theodore M. "Genres and Objects of Social Inquiry, from the Enlightenment to 1890." In *The Cambridge History of the Modern Social Sciences*, edited by Theodore M. Porter and Dorothy Ross, 13–40. Cambridge: Cambridge University Press, 2003.

———. "Precision and Trust: Early Victorian Insurance and the Politics of Calculation." In *The Values of Precision*, edited by M. Norton Wise, 173–97. Princeton, NJ: Princeton University Press, 1995.

Poundstone, William. *Prisoner's Dilemma: John von Neumann, Game Theory, and the Puzzle of the Bomb*. New York: Doubleday, 1992.

Powers, Willow Roberts. "The Harvard Study of Values: Mirror for Postwar Anthropology." *Journal of the History of the Behavioral Sciences* 36, no. 1 (2000): 15–29.

Putnam, Hilary. *Reason, Truth and History*. Cambridge: Cambridge University Press, 1981.

Quattrone, George A., and Amos Tversky. "Contrasting Psychological and Rational Analyses of Political Choice." *American Political Science Review* 82 (1988): 719–36.

———. "Self-Deception and the Voter's Illusion." *Journal of Personality and Social Psychology* 46 (1984): 237–48.

Ramsey, F. P. *The Foundations of Mathematics and Other Logical Essays*. London: Routledge and Kegan Paul, 1931.

RAND Corporation. *The RAND Corporation: The First Fifteen Years*. Santa Monica, CA: RAND Corporation, 1963.

Rapoport, Anatol. *Certainties and Doubts: A Philosophy of Life*. Toronto: Black Rose Books, 2000.

———. "Chicken à la Kahn." *Virginia Quarterly Review* 41 (1965): 370–89.

———. "Escape from Paradox." *Scientific American* (July 1967): 50–56.

———. "Lewis F. Richardson's Mathematical Theory of War." *Conflict Resolution* 1.3 (Sept 1957): 249–99.

———. "Rejoinder to Wohlstetter's Comments." In *Strategic Interaction and Conflict*, edited by Kathleen Archibald, 88–101. Berkeley: International Security Program, Institute of International Studies, University of California at Berkeley, 1966.

———. *Strategy and Conscience*. New York: Harper and Row, 1964.

Rapoport, Anatol, and Albert M. Chammah. *Prisoner's Dilemma: A Study in Conflict and Cooperation*. Ann Arbor: University of Michigan Press, 1965.

Rapoport, Anatol, and Melvin Guyer. "A Taxonomy of 2×2 Games." *General Systems* 11 (1966): 203–14.

Rapoport, Anatol, and Carol Orwant. "Experimental Games: A Review." *Behavioral Science* 7 (1962): 1–37.

Rawls, John. *A Theory of Justice*. Cambridge, MA: Harvard University Press, 1971.

———. "Kantian Constructivism in Moral Theory." *Journal of Philosophy* 77 (1980): 515–72.

———. *Political Liberalism*. Rev. ed. New York: Columbia University Press, 1996.

Rees, Mina. "The Mathematical Sciences and World War II." *American Mathematical Monthly* 87, no. 8 (1980): 607–21.

Reisberg, Daniel. *Cognition: Exploring the Science of the Mind*. New York: W. W. Norton, 1997.

Reisch, George. *How the Cold War Transformed Philosophy of Science: To the Icy Slopes of Logic*. Cambridge: Cambridge University Press, 2005.

Rey, Alain, ed. *Le Robert Dictionnaire historique de la langue française*. Paris: Dictionnaires Le Robert, 2000.

Riche de Prony, Gaspard. *Notices sur les grandes tables logarithmiques et trigonométriques, adaptées au nouveau système decimal*. Paris: Firmin Didot, 1824.

Rider, Robin E. "Operations Research and Game Theory—Early Connections." In *Toward a History of Game Theory*, edited Roy E. Weintraub, 225– 40. Durham, NC: Duke University Press, 1992.

Rips, Lance J., and S. L. Marcus. "Supposition and the Analysis of Conditional Sentences." In *Cognitive Processes in Comprehension*, edited by Marcel A. Just and Patricia A. Carpenter, 185–220. Hillsdale, NJ: Erlbaum, 1978.

Riskin, Jessica. "The Defecating Duck, or the Ambiguous Origins of Artificial Life." *Critical Inquiry* 29 (2003): 599–633.

Ritter, Joachim, Karlfried Gründer, and Gottfried Gabriel, eds. *Historisches Wörterbuch der Philosophie*, 13 vols. Basel: Schwabe, 1971–2007.

Rosenberg, David Alan. "The Origins of Overkill: Nuclear Weapons and American Strategy, 1945-1960." *International Security* 7 (1983): 3–71.

Rosenblueth, Arturo, Julian Bigelow, and Norbert Wiener. "Behavior, Purpose and Teleology." *Philosophy of Science* 10 (1943): 18–24.

Ross, Dorothy. "Changing Contours of the Social Science Disciplines." In *The Cambridge History of the Modern Social Sciences*, edited by Theodore M. Porter and Dorothy Ross, 205–37. Cambridge: Cambridge University Press, 2003.

Rothacker, Ernst. *Logik und Systematik der Geisteswissenschaften*. Bonn: H. Bouvier u. Co. Verlag, 1947.

Rothschild, Emma. "Condorcet and the Conflict of Values." *Historical Journal* 3 (1996): 677-701.

Rubinstein, Ariel. "Equilibrium in Supergames with the Overtaking Criterion." *Journal of Economic Theory* 21 (1979): 1-9.

Russell, Bertrand. *Common Sense and Nuclear Warfare*. London: George Allen & Unwin, 1959.

Sagan, Scott D., and Jeremi Suri. "The Madman Nuclear Alert: Secrecy, Signaling, and Safety in October 1969." *International Security* 27.4 (2003): 150-83.

Saint-Miel, Smaragdus of. *Commentary on the Rule of Saint Benedict*. Translated by David Barry. Kalamazoo, MI: Cistercian Publications, 2007.

Salvesen, Melvin. "The Institute of Management Sciences: A Prehistory and Commentary on the Occasion of TIMS' 40th Anniversary." *Interfaces* 27, no. 3 (1997): 74-85.

Samuels, Richard, Stephen Stich, and Michael Bishop. "Ending the Rationality Wars: How to Make Disputes about Human Rationality Disappear." In *Common Sense, Reasoning and Rationality*, edited by Renée Elio, 236-68. Oxford: Oxford University Press, 2002.

Sanford, George. *Military Rule in Poland: The Rebuilding of Communist Power, 1981-1983*. London: Croom Heim, 1986.

Sang, Edward. "Remarks on the Great Logarithmic and Trigonometrical Tables Computed in the Bureau de Cadastre under the Direction of M. Prony." *Proceedings of the Royal Society of Edinburgh* (1874-1875): 10.

Savage, Leonard J. *The Foundations of Statistics*. New York: John Wiley & Sons, 1954.

Schafer, Mark, and Scott Crichlow. "Antecedents of Groupthink: A Quantitative Study." *Journal of Conflict Resolution* 40 (1996): 415-35.

Schaffer, Simon. "Babbage's Intelligence: Calculating Engines and the Factory System." *Critical Inquiry* 21 (1994): 203-27.

———. "Enlightened Automata." In *The Sciences in Enlightened Europe*, edited by William Clark, Jan Golinski, and Simon Schaffer, 126-65. Chicago: University of Chicago Press, 1999.

———. "Genius in Romantic Natural Philosophy." In *Romanticism in the Sciences*, edited by Andrew Cunningham and Nicholas Jardine, 82-98. Cambridge: Cambridge University Press, 1990.

Schell, Emil D. "Application of the Univac to Air Force Programming." *Proceedings of the Fourth Annual Logistics Conference*, 1-7. Washington DC: Navy Logistics Research Project, 1953.

Schelling, Thomas C. "Discussion. First Session: The Concept of Rationality." In *Strategic Interaction and Conflict*, edited by Kathleen Archibald, 138-56. Berkeley: International Security Program, Institute of International Studies, University of California at Berkeley, 1966.

———. "Meteors, Mischief, and War." *Bulletin of the Atomic Scientists* 16 (1960): 292-300.

———. *Strategies of Commitment and Other Essays*. Cambridge, MA: Harvard University Press, 2006.

———. *The Strategy of Conflict*. Cambridge, MA: Harvard University Press, [1960] 1980.

———. "Uncertainty, Brinkmanship, and the Game of Chicken." In *Strategic Interaction and Conflict*, edited by Kathleen Archibald, 74-87. Berkeley: International Security Program, Institute of International Studies, University of California at Berkeley, 1966.

Schelling, Thomas C. "Autobiography." Nobelprize.org. 2012. http://www.nobelprize.org/nobel_prizes/economics/laureates/2005/schelling-autobio.html.

Selten, Reinhard. "What Is Bounded Rationality?" In *Bounded Rationality: The Adaptive*

Toolbox, edited by Gerd Gigerenzer and Reinhard Selten, 13–36. Cambridge, MA: MIT Press, 2001.

Semmel, Andrew K., and Dean Minix. "Small-Group Dynamics and Foreign Policy Decision-Making: An Experimental Approach." In *Psychological Models in International Politics,* edited by Lawrence S. Falkowski, 251–287. Boulder, CO: Westview Press, 1979.

Sent, Esther Mirjam. "Herbert A. Simon as a Cyborg Scientist." *Perspectives on Science* 8, no. 4 (2000): 380–406.

———. "Simplifying Herbert Simon." *History of Political Economy* 37, no. 2 (2005): 227–32.

Shubik, Martin. "Game Theory at Princeton, 1949–1955: A Personal Reminiscence." In *Toward a History of Game Theory,* edited by E. Roy Weintraub, 151–64. Durham, NC: Duke University Press, 1992.

Sibley, M. "The Rational versus the Reasonable." *Philosophical Review* 62 (1953): 554–60.

Sidgwick, Henry. *The Methods of Ethics.* London: Macmillan, 1874.

Simon, Herbert A. "A Behavioral Model of Rational Choice." *Quarterly Journal of Economics* 69, no. 1 (1955): 99–118.

———. "Dynamic Programming under Uncertainty with a Quadratic Criterion Function." *Econometrica* 24 (1956): 74–81.

———. "From Substantive to Procedural Rationality." In *Method and Appraisal in Economics,* edited by S. J. Latsis, 120–48. New York: Cambridge University Press, 1976.

———. "Invariants of Human Behavior." *Annual Review of Psychology* 41 (1990): 1–19.

———. *Models of Bounded Rationality.* Cambridge, MA: MIT Press, 1982.

———. *Models of Man: Social and Rational; Mathematical Essays on Rational Human Behavior in a Social Setting.* New York: Wiley, 1957.

———. "Notes on Two Approaches to the Production Rate Problem." CCDP Economics 2057 (1952). http://cowles.econ.yale.edu/P/ccdp/ec/e-2057.pdf.

———. "On How to Decide What to Do." *Bell Journal of Economics* 9 (1978): 494–507.

———. "Rational Decision Making in Business Organizations." *American Economic Review* 69 (1979): 493–513.

———. "Rationality as Process and as Product of Thought." *American Economic Review* 68 (1978): 1–16.

———. "Some Further Requirements of Bureaucratic Theory." In *Reader in Bureaucracy,* edited by Robert K. Merton, Alisa P. Gray, Barbara Hockey, and Hanan C. Selvin, 51–58. Glencoe, IL: Free Press, 1952.

———. "Theories of Bounded Rationality." In *Decision and Organization,* edited by C. B. McGuire and Roy Radner, 161–176. Amsterdam: North-Holland, 1972.

Simon, Herbert A., and Charles C. Holt. "The Control of Inventory and Production Rates; A Survey." ONR research memorandum no. 9. Pittsburgh, PA: Graduate School of Industrial Administration Carnegie Institute of Technology, 1954.

Simpson, Christopher, ed. *Universities and Empire: Money and Politics in the Social Sciences in the Cold War.* New York: New Press, 1998.

Sloan, Bill. *Brotherhood of Heroes: The Marines at Peleliu, 1944—The Bloodiest Battle of the Pacific War.* New York: Simon and Schuster, 2005.

Smith, Brewster. "The American Soldier and Its Critics: What Survives the Attack on Positivism?" *Social Psychology Quarterly* 47, no. 2 (1984): 192–98.

Smith, Bruce L. R. *The RAND Corporation: Case Study of a Nonprofit Advisory Corporation.* Cambridge, MA: Harvard University Press, 1966.

Snider, James G., and Charles E. Osgood, eds. *Semantic Differential Technique: A Sourcebook.* Chicago: Aldine, 1969.

Snyder, Jack L. "Rationality at the Brink: The Role of Cognitive Processes in Failures of Deterrence." *World Politics* 30 (1978): 345–65.

Spielman, Richard. "Crisis in Poland." *Foreign Policy* 49 (1982–1983): 20–36.

Spindler, George, Bert Gerow, and John Dodds. "Felix Manning Keesing, 1902–1961, Memorial Declaration." Stanford University Faculty Memorials. http://histsoc.stanford .edu/memorials.shtml.

Spiro, Melford. "The Problem of Aggression in a South Seas Culture." PhD diss., Northwestern University, 1950.

Stanovich, Keith E., and Richard F. West. "Evolutionary versus Instrumental Goals: How Evolutionary Psychology Misconceives Human Rationality." In *Evolution and the Psychology of Thinking: The Debate*, edited by David E. Over, 171–230. New York: Psychology Press, 2003.

Stein, Edward. *Without Good Reason: The Rationality Debate in Philosophy and Cognitive Science*. Oxford: Oxford University Press, 1996.

Stenning, Keith, and Michiel van Lambalgen. "The Natural History of Hypotheses about the Selection Task." In *Psychology of Reasoning*, edited by Ken I. Manktelow and Man C. Chung, 127–56. Hove & New York: Psychology Press, 2004.

Stern, Sheldon M. *The Week the World Stood Still: Inside the Secret Cuban Missile Crisis*. Stanford, CA: Stanford University Press, 2005.

Stigler, George. "The Costs of Subsistence." *Journal of Farm Economics* 27, no. 2 (May 1945): 303–14.

———. "The Development of Utility Theory, Parts I and II." *Journal of Political Economy* 58 (1950): 307–27, 373–96.

Stigler, Stephen M. *The History of Statistics: The Measurement of Uncertainty before 1900*. Cambridge, MA: Harvard University Press, 1986.

———. "Thomas Bayes's Bayesian Inference." *Journal of the Royal Statistical Society (A)* 145 (1982): 250–58.

Stouffer, Samuel et al. *Studies in Social Psychology in World War II*. Princeton, NJ: Princeton University Press, 1949.

Sturm, Thomas. *Kant und die Wissenschaften vom Menschen*. Paderborn: Mentis, 2009.

Suedfeld, Peter. "Cognitive Managers and Their Critics." *Political Psychology* 13 (1992): 435–543.

"Summary of Declassified Nuclear Stockpile Information." U.S. Department of Energy OpenNet. http://www.osti.gov/opennet/forms.jsp?formurl=document/press/pc26tab1 .html.

Sutherland, Stuart. *Irrationality*. London: Pinker & Martin, 1992.

Swade, Doron. *The Difference Engine: Charles Babbage and the Quest to Build the First Computer*. New York: Viking, 2001.

Tetlock, Philip E. "Correspondence and Coherence: Indicators of Good Judgment in World Politics." In *Thinking: Psychological Perspectives on Reasoning*, edited by David Hardman and Laura Macchi, 233–50. Chichester, UK: Wiley, 2003.

———. *Expert Political Judgment: How Good Is It? How Can We Know?* Princeton, NJ: Princeton University Press, 2005.

———. "Good Judgment in International Politics: Three Psychological Perspectives." *Political Psychology* 13 (1992): 517–39.

———. "Theory-Driven Reasoning about Plausible Pasts and Probable Futures in World Politics." In *Heuristics and Biases*, edited by Thomas Gilovich, Dale W. Griffin, and Daniel Kahneman, 749–62. Cambridge: Cambridge University Press, 2002.

Tetlock, Philip E., Charles B. McGuire, and Gregory Mitchell. "Psychological Perspectives on Nuclear Deterrence." *Annual Review of Psychology* 42 (1991): 239–76.

Thomas, Gerald William. "A Veteran Science: Operations Research and Anglo-American Scientific Cultures, 1940–1960." PhD diss., Harvard University, 2007.

Tresch, John. *The Romantic Machine: Utopian Science and Technology after Napoleon.* Chicago: University of Chicago Press, 2012.

Trivers, Robert. "The Evolution of Reciprocal Altruism." *Quarterly Review of Biology* 46 (1971): 35–57.

Trusteeship Agreement for the Former Japanese Mandated Islands, United Nations-United States, July 18, 1947, 61 Stat. 3301, Treaties and other International Acts Series. No. 1665.

Tucker, A. W., and R. D. Luce, eds. *Contributions to the Theory of Games.* Vol. 4. Princeton, NJ: Princeton University Press, 1959.

Tunner, William H. *Over the Hump.* Repr. ed. Washington: Office of Air Force History United States Air Force, 1985.

Turing, Alan M. "Computing Machinery and Intelligence." *Mind* 59 (1950): 433–60.

———. "On Computable Numbers, with an Application to the *Entscheidungsproblem.*" *Proceedings of the London Mathematical Society,* ser. 2, 42 (1936–1937): 230–65.

Tversky, Amos, and Daniel Kahnemann. "Advances in Prospect Theory: Cumulative Representation of Uncertainty." *Journal of Risk and Uncertainty* 5 (1992): 297–323.

———. "Belief in the Law of Small Numbers." *Psychological Bulletin* 2 (1971): 105–10.

———. "Extensional versus Intuitive Reasoning: The Conjunction Fallacy in Probability Judgment," *Psychological Review* 90 (1983): 293–315.

———. "Judgment under Uncertainty: Heuristics and Biases." *Science* 185 (1974): 1124–31.

———. "On the Reality of Cognitive Illusions." *Psychological Review* 103 (1996): 582–91.

———. "Rational Choice and the Framing of Decisions." *Journal of Business* 59 (1986): S251–78.

U.S. Air Force Planning Research Division Director of Program Standards and Cost Control Comptroller. *Scientific Planning Techniques: A Special Briefing for the Air Staff 5 August 1948.* Project SCOOP discussion papers, 1-DU. Washington DC: August 5, 1948.

U.S. Department of State, Foreign Policy Studies Branch, Division of Historical Policy Research. "The Berlin Crisis." Research project No. 171.Washington DC: Government Printing Office, 1948.

Van Heijenoort, Jean, ed. *From Frege to Gödel: A Sourcebook in Mathematical Logic, 1879–1931.* Cambridge, MMA.: Harvard University Press, 1967.

Vandenberg, General Hoyt S. "Air Force Letter No. 170–3, Comptroller Project SCOOP." Washington DC, October, 13, 1948, Air Force Historical Research Agency IRIS Number 01108313.

Vaucanson, Jacques. *Le mécanisme du flûteur automate.* Paris: Guerin, 1738.

Verba, Sidney. "Assumptions of Rationality and Non-Rationality in Models of the International System." *World Politics* 14 (1961): 93–117.

Vilkas, E., ed. *Uspekhi teorii igr: Trudy II Vsesoiuznoi konferentsii po teorii igr. Vil'nius 1971.* Vilnius, Lithuania: Mintis, 1973.

Von Neumann, John, and Oskar Morgenstern. *Theory of Games and Economic Behavior.* Princeton, NJ: Princeton University Press, 1944.

———. *Theory of Games and Economic Behavior*. 2nd ed. Princeton, NJ: Princeton University Press, 1953.

Vorob'ev, N. N. "Nauchnye itogi konferentsii," in *Uspekhi teorii igr*, edited by E. Vilkas, 7–13. Vilnius, Lithuania: Mintis,1973.

———. "Prilozheniia teorii igr (Metodologicheskii ocherk)," in *Uspekhi teorii igr*, edited by E. Vilkas, 249–83. Vilnius, Lithuania: Mintis,1973.

Vorob'ev, N. N. "Sovremennoe sostoianie teorii igr," in *Teoriia igr*, edited by N. N. Vorob'ev et al., 5–57. Erevan, Armenia: Izd. AN Armianskoi SSR, 1973.

Ware, Willis H. "RAND Contributions to the Development of Computing." http://www .rand.org/about/history/ware.html.

Wason, Peter C. "Realism and Rationality in the Selection Task." In *Thinking and Reasoning*, edited by Jonathan St. B. T. Evans, 44–75. London: Routledge & Kegan Paul, 1983.

———. "Reasoning." In *New Horizons in Psychology*, edited by Brian M. Foss, 135–51. Harmondsworth, UK: Penguin, 1966.

———. "Reasoning about a Rule." *Quarterly Journal of Experimental Psychology* 20 (1968): 273–81.

Weiner, M.G. "An Introduction to War Games." RAND paper P-1773, 1959.

Weintraub, E. Roy, ed. *Toward a History of Game Theory*. Durham, NC: Duke University Press, 1992.

Weisbord, Marvin. *Productive Workplaces Revisited. Dignity, Meaning, and Commmunity in the 21st Century*. San Francisco: Jossey-Bass 2004.

Weizenbaum, Joseph. *Computer Power and Human Reason: From Judgment to Calculation*, San Francisco: W. H. Freeman, 1976.

Welch, David A., and James G. Blight. "The Eleventh Hour of the Cuban Missile Crisis: An Introduction to the ExComm Transcripts." *International Security* 12 (Winter 1987–1988): 5–29.

White, Theodore H. "The Action-Intellectuals," with photographs by John Lonegard. *Life Magazine*, June 9, 1967, 43–76; June 16, 1967, 44–74B; June 23, 1967, 76–87.

Whitman, Howard. "How to Keep Out of Trouble." *Collier's Weekly*, September 25, 1948, 28–41.

Wise, Norton M. "The Gender of Automata in Victorian Science." In *Genesis Redux: Essays on the History and Philosophy of Artificial Life*, edited by Jessica Riskin, 163–95. Chicago: University of Chicago Press, 2007.

Wittgenstein, Ludwig. *Philosophical Investigations*. Translated by G. E. M. Anscombe. Englewood Cliffs, NJ: Prentice Hall, 1958.

Wohlstetter, Albert. "Comments on Rapoport's Paper: The Non-Strategic and the Non-Existent." In *Strategic Interaction and Conflict*, edited by Kathleen Archibald, 107–26. Berkeley: International Security Program, Institute of International Studies, University of California at Berkeley, 1966.

———. "The Delicate Balance of Terror." *Foreign Affairs* 37 (January 1959): 211–34.

Wood, Marshall K. "Research Program at Project SCOOP." Symposium on Linear Inequalities and Programming, Washington DC, June 14–16, 1951, Project SCOOP Manual no. 10, April 1, 1952.

Wood, Marshall K., and George B. Dantzig. "Programming of Interdependent Activities: I General Discussion." *Econometrica* 17 (1949): 193–99.

Wood, Marshall K., and Murray A. Geisler. "Development of Dynamic Models for Program Planning." In *Activity Analysis of Production and Allocation: Proceedings of a Confer-*

ence, Cowles Commission for Research in Economics Monograph No. 13. edited by Tjalling C. Koopmans, 89–215. New York: John Wiley & Sons, 1951.

———. "Machine Computation of Peacetime Program Objectives and Mobilization Programs." Project SCOOP No. 8, report prepared for Planning Research Division Director of Program Standards and Cost Control Comptroller, Headquarters U.S. Air Force. Washington DC, July 18, 1949.

Zarate, Robert, and Henry Sokolski, eds. *Nuclear Heuristics: Selected Writings of Albert and Roberta Wohlstetter*. Carlisle, PA: Strategic Studies Institute, U.S. Army War College, 2009.

ARCHIVAL SOURCES

Bales, Robert Freed. Papers. Rockefeller Archive Center, Sleepy Hollow, New York.

Condorcet (Dossier). Archives de l'Académie des Sciences, Paris.

Cornell-Peru Project Vicos Collection, Carl A. Kroch Library, Cornell University, Division of Rare and Manuscript Collections, Ithaca, New York.

Flood, Merrill M. Papers. Bentley Historical Library, University of Michigan.

George Price Papers, W. D. Hamilton Archive, British Library.

Hamilton, W. D. Archive. British Library, London.

Herbert A. Simon Collection. Carnegie Mellon University Archives, Pittsburgh, Pennsylvania.

Office of Naval Intelligence Monograph File, POA

Oskar Morgenstern Papers. Duke University Special Collections, Durham, North Carolina.

INDEX

Made in the USA
Las Vegas, NV
17 February 2022